Hypertension: Insights

Hypertension: Insights

Shivsanjeevi Sripathi and Prerna Pandey

Delve
PUBLISHING

www.delvepublishing.com

Hypertension: Insights

Shivsanjeevi Sripathi and Prerna Pandey

Delve Publishing
224 Shoreacres Road
Burlington, ON L7L 2H2
Canada
www.delvepublishing.com

Email: orders@arclereducation.com

ISBN: 978-1-77469-163-2 (Hardcover)

ABOUT THE AUTHORS

ShivSanjeevi Sripathi completed his Masters in Biotechnology from Mumbai University in 2008. He was awarded for academic excellence in both his Bachelors and Masters for securing second rank in Mumbai University in 2006 and first rank in his college: Kishinchand chellaram College. For his Masters he secured first rank in his college KET's V.G.Vaze College. He qualified CSIR and NET and and TOEFL in September 2008. He then worked on a stem cell project at the Specialized Centre for Cell Based Therapy (SCCT),KEM Hospital at Mumbai on a project entitled, : Isolation & detection of stem cells from Human Umbilical cord/ amniotic membrane" following which he worked at Junior Research Fellow at Microbiology & Cell Biology Department, Indian Institute of Science, Bangalore on cloning of cell wall genes and transcription factors in E.coli & M.smegmatis. As a writer, he has authored and co-authored 35 books on various aspects of biology such as bionics, molecular wires, cloning, hypertension, the epidemics of the 21st century, handling depression, camouflage, hygiene, immunology and many more with international publishers. He loves to read and share on interesting aspects of life sciences in books. In his free time he loves to travel and explore and give talks on spirituality, ancient customs and traditions.

Dr Prerna Pandey, a biotechnologist with several years of wet lab research experience, worked at International Center for Genetic Engineering and Biotechnology, New Delhi. Her field of research in PhD included isolation and molecular characterization of geminiviruses, genome sequencing, gene annotation, and gene silencing using the RNA interference technology. She has also worked at Transasia Biomedicals and Advance Enzyme Technologies as a scientist. Prerna has published papers in peer-reviewed journals, and has submitted a number of annotated Geminiviral genome sequences in the GenBank, including two novel ones. She has also completed her editing and proof-reading courses from Society for the promotion of Editors and Proof-readers, UK and now works as a freelance scientific editor and writer. When Prerna is not busy with her assignments, she loves to explore historical places.

TABLE OF CONTENTS

Acknowledgement .... *ix*

Preface .... *xi*

Chapter 1 Insights into Hypertension: A Slew of Studies 1

1.1 A Global Picture 2

1.2 Staging the Condition 5

1.3 Masked and White-Collar Hypertension 10

1.4 Socio-Economic Factors and Hypertension 13

1.5 Variations 17

1.6 A Walk through History 20

1.7 Barriers 23

1.8 Recommendation Guidelines 32

1.9 Smartphones and Hypertension Monitoring 34

1.10 Patient and Physician Perspectives 36

1.11 References 43

Chapter 2 Hypertension in Children and Adolescents: The Cases and the Solutions 47

2.1 The Normal Values 48

2.2 What Are The Figures? A Global Snapshot from the US to Africa 53

2.3 Grading and Diagnosis 60

2.4 The Quest for Links and Factors Involved 65

2.5 A neonatal Link 86

2.6 Hypertension and Sleep 89

2.7 The Effect of Seasons 96

2.8 Physical Activity 97

2.9 Final Points 104

2.10 References: 107

Chapter 3 **Controlling Hypertension and Research Insights: A Glimpse 111**
3.1 Diet 112
3.2 miRNAs of the Gut-Liver Axis 133
3.3 References 142

Chapter 4 **Changes in the Body Reported Due to Hypertension-Case Studies on the Brain, Depression, Quality of Life and Diabetes 145**
4.1 The Brain 146
4.2 Hypertensive vs. Non-Hypertensive 149
4.3 Pre-hypertensive vs. Normotensive 150
4.4 Medicated vs. Non-medicated 151
4.5 Hypertensive vs. Non-hypertensive 153
4.6 Pre-hypertensive vs. Normotensive 153
4.7 Medicated vs. Non-medicated 153
4.8 The Blues: Depression 166
4.9 Baseline vs. Post-PMR Scores for Anxiety, Depression, and Quality of Life (QOL) with Six-Minute Walk Distance 187
4.10 Diabetes and Hypertension 205
4.11 Final Points 215
4.12 References 216

Chapter 5 **Hypertension and COVID-19: What does Research say? 219**
5.1 The Extent of Damage 220
5.2 Hypertension Sets the Stage 222
5.3 Perspectives of Hypertensive Patients 248
5.4 References 261

Index 263

ACKNOWLEDGEMENTS

Thank you my dear parents Sripathi and Nirmala for all the blessings and support. I am grateful to Dr. Prerna Pandey and the publisher for this opportunity.

Dedicated to the late Swami Omkarananda and each person out there who is under or has felt "pressure". Stay strong!

सर्वे भवन्तु सुखिनः सर्वे सन्तु निरामयाः।

सर्वे भद्राणि पश्यन्तु मा कश्चिद्दुःखभाग् भवेत्।।

May all be happy; May all be free from disease; May all see good; May none undergo sorrow

-Ancient Indian quote

PREFACE

Do you feel under "pressure" with changing seasons? Read on to know-how!

Blood pressure is the force of the blood against the walls of the arteries of the pumped blood. The systolic pressure is the pressure on the blood vessels when the heart beats while the diastolic pressure is the force of blood in the blood vessels when the heart relaxes between the heartbeats. According to the US FDA, we call a blood pressure of 120/80 as "normal". Stage 1 hypertension is at 130/80 and 140/90 or higher is stage 2 hypertension.

So why this book?

The world is facing the COVID-19 pandemic that is affecting lives in multiple dimensions. Apart from the physical, emotional and financial damage wreaked, of prime importance is care for non-communicable diseases such as diabetes and hypertension.

The symptoms range from buzzing in the ears to nose bleeding to morning headaches to confusion, muscle tremors, nausea and fatigue, hypertension is described as the "*silent killer*" often showing no distinct symptoms or warning signs. The condition is estimated to affect **1.13 billion people** across the globe with two-thirds in low/middle-income countries. The only way we can know if we have hypertension is by getting it measured by a healthcare professional. Hypertension can increase the risk of a slew of diseases such as heart disease, kidney disease, and stroke among others.

"One way to get high blood pressure is to go mountain climbing over molehills".

- Earl Wilson

This book presents the prevalence of hypertension along with several impacts on the body as outlined by the latest research. This is inclusive of cognitive decline, its association with depression and diabetes as presented by case studies and presents plausible solutions presented by scientists to help tide over the conditions.

"Adolescence is a new birth, for the higher and more completely human traits are now born".

-G. Stanley Hall

Along with the multiple changes that occur physically, mentally and emotionally during adolescence, another issue is hypertension. The CDC reported that one of seven US youth (12-19 years old) had elevated blood pressure on the basis of the 2017 AAP Clinical Practice Guidelines. Research for example, (Anyaegbu & Dharnidharka, *Pediatric clinics of North America*, 2014) points out a prevalence of 1-5% of hypertension in children. This book presents research-backed data as to the trend in hypertension in children and adolescents along with associated factors such as sleep and even the impact of seasons! This is followed up with recommendations from researchers.

Another USP of this book is the connection between COVID-19 infection and hypertension. Research-backed prevalence of data along with affected patient perspectives and other risk factors can help us to tackle this issue rather than panicking.

Other aspects covered range from the connection of hypertension with "status", discrepancies in the treatment observed in low-income vs. high-income countries and finally, perspectives from both patients and physicians are also presented to help gain an understanding of the condition. The utility of the DASH diet and the importance of micronutrients in controlling hypertension are also discussed. New aspects such as the role of miRNAs of the gut-lung axis and the impact of gut microbes have also been touched upon.

Overall, this book presents the current scenario of hypertension along with new advances and current gaps to help us to tackle the "silent killer".

"Good health is not something we can buy. However, it can be an extremely valuable savings account"

-Anne Wilson Schaef

Author's note: The contents of the book are not diagnostic or suggestive of treatments. Please consult a certified healthcare professional for diagnosis and treatment.

Chapter 1

Insights into Hypertension: A Slew of Studies

Shivsanjeevi Sripathi

"The secret of health for both mind and body is not to mourn for the past, not to worry about the future, or not to anticipate troubles, but to live the present moment wisely and earnestly."

-Buddha

CONTENTS

1.1 A Global Picture 2
1.2 Staging the Condition 5
1.3 Masked and White-Collar Hypertension 10
1.4 Socio-Economic Factors and Hypertension 13
1.5 Variations 17
1.6 A Walk through History 20
1.7 Barriers 23
1.8 Recommendation Guidelines 32
1.9 Smartphones and Hypertension Monitoring 34
1.10 Patient and Physician Perspectives 36
1.11 References 43

1.1 A GLOBAL PICTURE

The prevalence of hypertension was scrutinized in a 2020-published research article by Zeng and team in the *Annals of Global Health*. Based on the WHO data of hypertension by sex across 182 countries, the researchers linked this trend with the Human development index (HDI) as estimated by the United Nations Development Programme (UNDP) both for the year 2015. This index gauges the burden by encompassing life expectancy (health), schooling time (education) and the gross national income per capita (standard of living).

	N (missing)	Min	Max	Percentiles			Mean	SD
				25	50	75		
Prevalence (%)								
Male	182(0)	15.0	45.0	21.0	23.0	28.0	25.0	5.9
Female	182(0)	9.0	38.0	19.0	21.0	24.0	21.7	5.0
Total	182(0)	13.0	41.0	20.0	22.0	25.0	23.4	5.2
HDI	182(0)	0.352	0.949	0.579	0.733	0.811	0.700	0.153
Life expectancy at birth	182(0)	50.1	83.7	66.0	73.5	77.0	71.5	8.1
Expected years of schooling	182(0)	5.0	20.4	10.9	13.2	15.0	13.0	2.8
Mean years of schooling	182(0)	1.4	13.4	6.1	8.7	11.1	8.4	3.1
GNI per capita	182(0)	587.0	129916.0	3433.3	10492.5	23517.0	17072.3	18601.5
Alcohol (%)								
Male	180(2)	0.0	25.0	4.0	11.0	16.0	10.4	6.6
Female	180(2)	0.0	7.0	1.0	2.0	3.0	2.2	1.8
Total	180(2)	0.0	15.0	2.3	6.0	9.0	6.2	4.0
Physical inactivity (%)								
Male	155(27)	4.0	60.0	16.0	24.0	32.0	23.7	10.4
Female	155(27)	6.0	73.0	23.0	33.0	42.0	32.7	13.2
Total	155(27)	5.0	65.0	19.0	29.0	37.0	28.2	11.4
Salt/sodium intake (%)								
Male	180(2)	4.0	16.0	7.0	9.0	11.0	9.2	2.5
Female	180(2)	4.0	14.0	7.0	8.5	10.0	8.4	2.2
Total	180(2)	4.0	15.0	7.0	9.0	10.0	8.8	2.3
Tobacco use (%)								
Male	143(39)	1.0	77.0	21.0	30.0	41.0	32.1	14.0
Female	144(38)	0.0	44.0	2.0	6.0	18.0	9.8	9.5
Total	142(40)	4.0	47.0	14.0	21.0	27.0	21.2	9.1

Hypertension ranged from 13.0% to 41.0% in the 182 countries with the HDI between 0.352 and 0.949. Table 1 above represents the results obtained (Zeng, Z., Chen, J., Xiao, C., & Chen, W. (2020), A Global View on Prevalence of Hypertension and Human Develop Index. *Annals of global health*, *86*(1), 67. https://doi.org/10.5334/aogh.2591 © 2020 The Author(s): *This is an open-access article distributed under the terms of the Creative Commons Attribution 4.0 International License (CC-BY 4.0), which permits unrestricted use, distribution, and reproduction in any medium, provided the original author and source are credited. See http://creativecommons.org/licenses/by/4.0/*):

From the hypertension prevalence and HDI emerged positive correlations for the entire population and males as outlined by Spearman's correlations analysis.

Table 2 below depicts this analysis:: ** Correlation is significant at the 0.01 level (2-tailed). * Correlation is significant at the 0.05 level (2-tailed) (Zeng, Z., Chen, J., Xiao, C., & Chen, W. (2020), A Global View on Prevalence of Hypertension and Human Develop Index. *Annals of global health*, *86*(1), 67. https://doi.org/10.5334/aogh.2591 © 2020 The Author(s): *This is an open-access article distributed under the terms of the Creative Commons Attribution 4.0 International License (CC-BY 4.0), which permits unrestricted use, distribution, and reproduction in any medium, provided the original author and source are credited. See http://creativecommons.org/licenses/by/4.0/*):

Factor	**Spearman's rho**	**Male**	**Female**	**Total**
HDI	Correlation Coefficient	0.385**	–0.017	0.215**
	P	<0.001	0.815	0.004
	N	182	182	182
Life expectancy at birth	Correlation Coefficient	0.314**	–0.077	0.150*
	P	<0.001	0.299	0.043
	N	182	182	182

Expected years of schooling	Correlation Coefficient	0.406**	0.065	0.262**
	P	<0.001	0.385	<0.001
	N	182	182	182
Mean years of schooling	Correlation Coefficient	0.414**	0.062	0.262**
	P	<0.001	0.403	<0.001
	N	182	182	182
GNI per capita	Correlation Coefficient	0.319**	–0.078	0.152*
	P	<0.001	0.296	0.041
	N	182	182	182
Alcohol rate	Correlation Coefficient	0.517**	0.313**	0.456**
	P	<0.001	0.000	<0.001
	N	180	180	180
Physical inactivity rate	Correlation Coefficient	0.262**	–0.106	0.086
	P	0.001	0.189	0.290
	N	155	155	155
Salt/sodium intake rate	Correlation Coefficient	0.273**	0.177*	0.223**
	P	<0.001	0.018	0.003
	N	180	180	180
Tobacco use rate	Correlation Coefficient	0.254**	0.251**	0.371**
	P	0.002	0.002	<0.001
	N	143	144	142

Scatter plots outlined a non-linear relationship between hypertension and HID with a spike in hypertension with an increase from the HDI from medium to high (this classification is on the basis of the UNDP: low (HDI < 0.55), medium (0.55 < HDI < 0.70), high (0.70 < HDI < 0.80), and very high (0.80 < HDI < 1.00).

1.2 STAGING THE CONDITION

The following section has been taken verbatim from open-access journals as the definitions involve technical definitions:

On the basis of the 2018 ESC/ESH Guidelines for the management of arterial hypertension applicable to 16 years and above (Iqbal AM, Jamal: Bookshelf ID: NBK539859PMID: 30969681: Copyright © 2021, StatPearls Publishing LLC: *This book is distributed under the terms of the Creative Commons Attribution 4.0 International License (http://creativecommons.org/licenses/by/4.0/), which permits use, duplication, adaptation, distribution, and reproduction in any medium or format, as long as you give appropriate credit to the original author(s) and the source, a link is provided to the Creative Commons license, and any changes made are indicated*):

- "Optimal: SBP less than 120mmHg and DBP less than 80mmHg
- Normal: SBP 120 to 129mmHg and/or DBP 80 to 84mmHg
- High normal: SBP 130 to 139mmHg and/or DBP 85 to 89mmHg
- Grade 1 hypertension: SBP 140 to 159mmHg and/or DBP 90 to 99mmHg
- Grade 2 hypertension: SBP 160 to 179mmHg and/or DBP 100 to 109mmHg
- Grade 3 hypertension: SBP greater than or equal to 180mmHg and/or DBP greater than or equal to 110mmHg
- Isolated systolic hypertension: SBP greater than or equal to 140mmHg and DBP less than 90mmHg (further classified into Grades as per above ranges of SBP)
- For ambulatory BP measurements (ABPM) and home BP measurements (HBPM):
- Daytime (or awake) mean SBP greater than or equal to 135mmHg and/or DBP greater than or equal to 85mmHg
- Night-time (or asleep) mean SBP greater than or equal to 120mmHg and/or DBP greater than or equal to 70mmHg
- 24 hr mean SBP greater than or equal to 130mmHg and/or DBP greater than or equal to 80mmHg
- Home BP mean SBP greater than or equal to 135mmHg and/or DBP greater than or equal to 85mmHg"

On the basis of the American College of Cardiology/American Heart Association Task Force on Clinical Practice Guidelines for 20 years of age and above:

- "Normal: SBP less than 120 and DBP less than 80mmHg;
- Elevated: SBP 120 to 129 and DBP less than 80mmHg;
- Stage 1 hypertension: SBP 130 to 139 or DBP 80 to 89mmHg;
- Stage 2 hypertension: SBP greater than or equal to 140 mmHg or greater than or equal to 90 mmHg.
- White coat hypertension is an office BP 130/80 mmHg or more but less than 160/100mmHg which comes down to 130/80mmHg or less after at least 3 months of anti-hypertensive therapy. Ambulatory or home blood pressure measurement is usually necessary for this diagnosis.
- Masked hypertension is an elevated office systolic BP 120 to 129mmHg, and diastolic BP less than 80mmHg but raised BP on ambulatory or home measurements (130/80mmHg or more)."

On the basis of the Seventh Report of the Joint National Committee on Prevention, Detection, Evaluation, and Treatment of High Blood Pressure: the JNC 7 report for those over 18 years of age:

- "Normal: SBP less than 120mmHg and DBP less than 80mmHg
- Pre-Hypertension: SBP 120 to 139mmHg and DBP 80 to 89mmHg
- Stage 1 Hypertension: SBP 140 to 159mmHg and DBP 90 to 99mmHg
- Stage 2 Hypertension: SBP greater than or equal to 160mmHg and DBP greater than or equal to 100mmHg"

(Iqbal AM, Jamal, 2020)

The **classification** of blood pressure: systolic blood pressure (SBP) and diastolic blood pressure (DBP) is presented below in table 3 (Majid Nour and Kemal Polat, 2020: *Copyright © 2020 Majid Nour and Kemal Polat. This is an open access article distributed under the Creative Commons Attribution License, which permits unrestricted use, distribution, and reproduction in any medium, provided the original work is properly cited*):

Blood pressure classification	SBP (mmHg)	DBP (mmHg)
Normal	<120	And <80
Prehypertension	120–139	Or 80–89
Stage-1 hypertension	140–159	Or 90–99
Stage-2 hypertension	≥160	Or ≥100

Table 4 below summarizes the SBP and DBP values for blood pressure based on the hypertension guidelines of the European society of hypertension (ESH) and the European society of cardiology (ESC) (Majid Nour and Kemal Polat, 2020: *Copyright © 2020 Majid Nour and Kemal Polat. This is an open access article distributed under the Creative Commons Attribution License, which permits unrestricted use, distribution, and reproduction in any medium, provided the original work is properly cited*):

Blood pressure type	**SBP**	**DBP**
Classification	mmHg	mmHg
Optimal	<120	<80
Normal	120–129	80–84
High normal	130–139	85–89
Grade 1 hypertension	140–159	90–99
Grade 2 hypertension	160–179	100–109

On the basis of the final English version of the 2018 Chinese Guidelines for Prevention and Treatment of Hypertension (Revised Edition), Liu presented a snapshot of comparison of these guidelines with those of other countries as depicted below in **table 5** (Liu, 2018: *Open Access: This article is licensed under a Creative Commons Attribution 4.0 International License, which permits use, sharing, adaptation, distribution and reproduction in any medium or format, as long as you give appropriate credit to the original author(s) and the source, provide a link to the Creative Commons licence, and indicate if changes were made. The images or other third party material* *in this article are included in the article's Creative Commons licence, unless indicated otherwise in a credit line to the material. If material is not included in the article's Creative Commons licence and your intended use is not permitted by statutory regulation or exceeds the permitted use, you will need to obtain permission directly from the copyright holder. To view a copy of this licence, visit http://creativecommons.org/licenses/by/4.0/. The Creative Commons Public Domain Dedication waiver (http://creativecommons.org/publicdomain/zero/1.0/) applies to the data made available in this article, unless otherwise stated in a credit line to the data*):

Key: ACC American College of Cardiology, *AHA* American Heart Association, *BP* blood pressure, *CHL* Chinese Hypertension League, *DBP* diastolic BP, *ESC* European Society of Cardiology, *ESH* European Society of Hypertension, *ISH* isolated systolic hypertension, *JSH* Japanese Society of Hypertension, *KSH* Korean Society of Hypertension, *NA* not available, *SBP* systolic BP; [a] DBP: 80–84 mmHg; [b] DBP: 85–89 mmHg.

BP category (mmHg)	**CHL 2018 [1]**	**KSH 2018 [7,8,9]**	**JSH 2019 [10]**	**AHA/ACC 2017 [2]**	**ESC/ESH 2018 [3]**
SBP< 120 and DBP< 80	Normal	Normal	Normal	Normal	Optimal
SBP: 120–129 and DBP< 80	High normal	Elevated	High normal	Elevated	Normal[a]
SBP: 130–139 and (or) DBP: 80–89		Prehyper-tension	Elevated	Grade 1 hy-pertension	High normal[b]
SBP: 140–159 and (or) DBP: 90–99	Grade 1 hyper-tension	Grade 1 hy-pertension	Grade 1 hyperten-sion	Grade 2 hy-pertension	Grade 1 hy-pertension
SBP: 160–179 and (or) DBP: 100–109	Grade 2 hyper-tension	Grade 2 hy-pertension	Grade 2 hyperten-sion		Grade 2 hy-pertension
SBP≥ 180 and (or) DBP≥ 110	Grade 3 hyper-tension		Grade 3 hyperten-sion		Grade 3 hy-pertension
SBP≥ 140 and DBP< 90	ISH	ISH	ISH	NA	ISH

Further, targets of blood pressure across populations are presented below in table 6 (Liu, 2018: *Open Access: This article is licensed under a Creative Commons Attribution 4.0 International License, which permits use, sharing, adaptation, distribution and reproduction in any medium or format, as long as you give appropriate credit to the original author(s) and*

Key:

"ACC American College of Cardiology, AHA American Heart Association, BP blood pressure, CAD coronary artery disease, CHL Chinese Hypertension League, CKD chronic kidney disease, DM diabetes mellitus, ESC European Society of Cardiology, ESH European Society of Hypertension, HFrEF heart failure with reduced ejection fraction, JSH Japanese Society of Hypertension, KSH Korean Society of Hypertension, NA not available.

[a] <130/80 mmHg, if tolerable or high risk

[b] <140/90 mmHg, if tolerable

[c] <130/80 mmHg, if high risk.

[d] <130/80 mmHg, with lacunar infarction.

[e] <140/90 mmHg for bilateral cervical arteries stenosis, main cerebral artery occlusion or unevaluated.

[f] 130–139/70–79 mmHg for people aged ≥65 years"

	CHL 2018 [1]	**KSH 2018 [7,8,9]**	**JSH 2019 [10]**	**AHA/ACC 2017 [2]**	**ESC/ESH 2018 [3]**
Young & middle-aged adults	<140/90 mmHg[a]	<140/90 mmHg[c]	<130/80 mmHg	<130/80 mmHg	120–130/70–79 mmHg
Elderly	65-79y <150/90 mmHg[b]	≥65y <140/90 mmHg	≥75y <140/90 mmHg	≥65y <130/80 mmHg	65–79y 130–139/70–79 mmHg
	≥80y <150/90 mmHg				≥80y 130–139/70–79 mmHg

DM	<130/80 mmHg	<140/85 mmHg [c]	<130/80 mmHg	<130/80 mmHg	120–130/70–79 mmHg [f]
CKD without proteinuria	<140/90 mmHg	<140/90 mmHg	<140/90 mmHg	<130/80 mmHg	130–139/70–79 mmHg
CKD with proteinuria	<130/80 mmHg	<130/80 mmHg	<130/80 mmHg	<130/80 mmHg	130–139/70–79 mmHg
Secondary prevention of stroke	<140/90 mmHg	<140/90 mmHg [d]	<130/80 mmHg [e]	<130/80 mmHg	120–130/70–79 mmHg [f]
CAD	<140/90 mmHg [a]	<130/80 mmHg	<130/80 mmHg	<130/80 mmHg	120–130/70–79 mmHg [f]
HFrEF	<130/80 mmHg	<130/80 mmHg	NA	<130/80 mmHg	NA

1.3 MASKED AND WHITE-COLLAR HYPERTENSION

The quantification of blood pressure outside a clinic has resulted in the following types:

- Sustained normotension (individuals with normal BP at office and out-of-office)
- Sustained hypertension (individuals with hypertension)
- White-coat hypertension (individuals with office hypertension and normal out-of-office BP),
- Masked hypertension (individuals with out-of-office hypertension and normal office BP)

(Alwan et al, 2019)

A population-based cross-sectional study published in 2014 scrutinized the factors associated with masked hypertension and white-coat hypertension employing SKIPOGH (Swiss Kidney Project on Genes in Hypertension) data, part of the EPOGH study (European Project on Genes in Hypertension). Analyses of 652 participants (mean age= 48 (±18) years) revealed that office hypertension was seen in 15.0% of individuals (an office BP ≥140/90 mmHg) while daytime ambulatory hypertension was seen in 28.2% (daytime ambulatory BP ≥135/85 mmHg). Further, Alwan and team (2019) reported the highest 24-hour BP values in sustained hypertension (SH) and masked hypertension (MH) while the highest office BP levels were observed in white-coat hypertension (WCH). Analyses revealed positive associations between age and overweight/obesity with masked hypertension while age demonstrated an association with white-coat hypertension (WCH).

Figure 1 below is illustrative of the factors linked to ***TOP***: masked hypertension (© 2014 Alwan et al. *This is an open-access article distributed under the terms of the Creative Commons Attribution License, which permits unrestricted use, distribution, and reproduction in any medium, provided the original author and source are credited*):

		Univariate				Age, sex, centre, BMI-adjusted			
	n	OR	95% CI		P	OR	95% CI		P
Age (years)		1.03	1.01	1.04	<0.001	1.02	1.00	1.03	0.012
Sex (female)	43	0.56	0.36	0.88	0.011	0.63	0.39	1.02	0.058
High-normal office BP	42	8.17	4.75	14.04	<0.001	6.68	3.72	12.02	<0.001
Smoking (yes)	20	0.74	0.43	1.27	0.271	0.78	0.44	1.40	0.404
Alcohol consumption (yes)	67	1.21	0.77	1.90	0.412	1.01	0.61	1.66	0.981
Normal-weight	40								
Overweight	45	3.54	2.10	5.99	<0.001	2.99	1.74	5.15	<0.001
Obese	18	4.26	2.09	8.66	<0.001	3.63	1.73	7.60	0.001
Physical activity (hours/week)		0.93	0.86	0.99	0.035	0.92	0.85	0.99	0.023
Education (up to primary)	9								
Secondary	53	1.47	0.68	3.20	0.329	1.48	0.63	3.48	0.364
Higher education	41	1.58	0.71	3.53	0.261	2.02	0.83	4.91	0.121
Family history of hypertension	48	0.98	0.63	1.51	0.921	0.81	0.50	1.31	0.391
Family history of MI* or stroke	35	1.82	1.14	2.93	0.013	1.49	0.87	2.57	0.150

*MI: myocardial infarction. Analyses included people with masked hypertension and people with normotension as a comparison group. Total sample size for this analysis is 489 participants.
doi:10.1371/journal.pone.0092522.t002

BOTTOM: white-coat hypertension and associated factors (© 2014 Alwan et al. *This is an open-access article distributed under the terms of the Creative Commons Attribution License, which permits unrestricted use, distribution, and reproduction in any medium, provided the original author and source are credited*):

		Univariate				Age, sex, centre, BMI-adjusted			
Variable	n	OR	95% CI		P	OR	95% CI		P
Age (years)		1.07	1.04	1.11	<0.001	1.07	1.03	1.11	<0.001
Sex (female)	7	0.55	0.21	1.48	0.236	0.61	0.21	1.76	0.364
Alcohol consumption (yes)	11	1.19	0.43	3.29	0.736	0.95	0.31	2.95	0.935
Normal-weight	6								
Overweight	8	4.14	1.40	12.25	0.010	2.43	0.73	8.05	0.146
Obese	3	4.66	1.11	19.61	0.036	2.96	0.60	14.56	0.181
Physical activity (hours/week)		0.97	0.85	1.11	0.674	0.91	0.79	1.06	0.230
Education (up to primary)	3								
Secondary	8	0.67	0.17	2.61	0.560	0.85	0.19	3.89	0.834
Higher education	6	0.70	0.17	2.89	0.616	1.24	0.24	6.31	0.799
Family history of hypertension	11	2.05	0.75	5.67	0.164	2.15	0.65	7.15	0.213
Family history of MI* or stroke	7	2.48	0.92	6.71	0.074	1.32	0.35	4.94	0.678

*MI: myocardial infarction. Analyses included people with white-coat hypertension and people with normotension as a comparison group. Total sample size for this analysis is 403 participants.
doi:10.1371/journal.pone.0092522.t003

The persistence of masked hypertension was evaluated in white-collar workers from three public insurance institutions by Xavier Trudel, Alain Milot, Chantal Brisson in *the International Journal of Hypertension* (2013).

Analyses of 1669 participants revealed that masked hypertension was higher in men at 17.7% as opposed to 11.1% in ladies. This hypertension was associated more with BMI, intake of alcohol and family CVD (cardiovascular disease) history. After 3 years, 25% of masked hypertension seen in 126 men and 27% of masked hypertension in 106 women showed progression into sustained hypertension.

The persistence trend of masked hypertension reported by Xavier Trudel, Alain Milot, Chantal Brisson is shown below by the researchers in figure 2 (Copyright © 2013 Xavier Trudel et al. *This is an open access article distributed under the Creative Commons Attribution License, which permits unrestricted use, distribution, and reproduction in any medium, provided the original work is properly cited*)

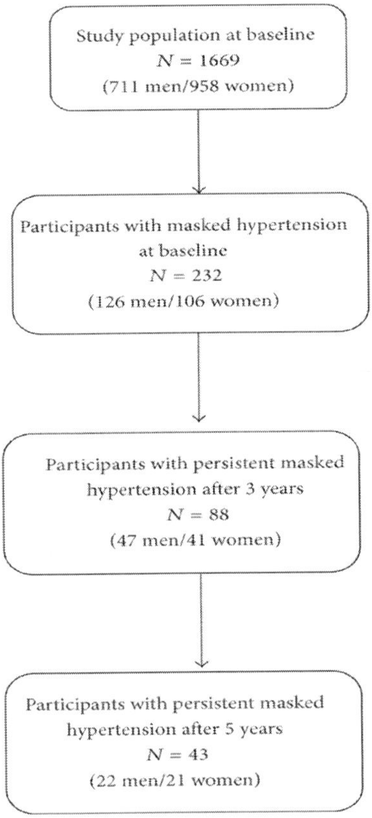

Overall, the researchers concluded that progression to sustained hypertension was seen in one-third of those with masked hypertension while this masking was persistent in one of five after five years (Trudel et al, 2013). This becomes pertinent given that the risk of stroke was higher in a cohort study of 4261 outpatients vs. those with controlled BP (hazard ratio, 2.77; 95% CI, 1.20-6.37) (Fujiwara et al, 2018).

1.4 SOCIO-ECONOMIC FACTORS AND HYPERTENSION

"Solitude is fine but you need someone to tell that solitude is fine."

— Honoré de Balzac

A term described as complicated is the socioeconomic status (SES) that is inclusive of education, income and where one lives that has been linked to predicting behavior and is a risk factor for cardiovascular disease. A team of researchers led by Sarah Cuschieri (2017) explored socioeconomic factors associated with 1189 individuals with hypertension in Malta. The differences in the various factors that influence hypertension are demonstrated by the researchers below (Copyright © 2017 Sarah Cuschieri et al. *This is an open access article distributed under the Creative Commons Attribution License, which permits unrestricted use, distribution, and reproduction in any medium, provided the original work is properly cited*):

		Hyper-tensive (total)	*Normo-tensive*	*p-value*	*Hypertensive*		*p-value*
					Male (729)	Female (529)	
Locality	Southern Harbour	585	215	**0.0001**	124	91	**0.0001**
	Northern Harbour	731	340		195	145	
	Southeastern	382	220		122	97	
	Western	360	186		109	77	
	Northern	423	163		91	72	
	Gozo	208	134		87	46	
Education	Low educational level	361	273	**0.0001**	160	201	**0.0001**
	Medium educational level	497	1034		284	213	
	High educational level	399	1382		285	114	

Employment status	Employed	625	1898	**0.0001**	473	153	**0.0001**
	Not in employment	632	791		256	376	
Smoking habit	Smoking	268	691	**0.003**	192	76	**0.001**
	Nonsmoking	990	1999		537	453	
Alcohol consumption	Alcohol consumption	1487	625	**0.001**	458	167	**0.0001**
	No-alcohol consumption	1203	633		271	361	
Physical activity	No activity	290	98	**0.015**	45	53	0.084
	Walk	522	230		138	91	
	Moderate activity	1585	783		458	325	
	Vigorous activity	292	147		88	59	

A 2019-published article in the International Journal of Hypertension by Xinyi Wu and Zhonghua Wang scrutinized the socioeconomic factors and hypertension in middle-aged and elderly people employing data from the China Health and Retirement Longitudinal Study (CHARLS). Analyses revealed the impact of being male, older age, obesity, having diabetes and lower social support as important factors associated with hypertension.
Table 7 below documents the results of the factors associated with hypertension with evident differences in bold (Copyright © 2019 Xinyi Wu and Zhonghua Wang. *This is an open access article distributed under the Creative Commons Attribution License, which permits unrestricted use, distribution, and reproduction in any medium, provided the original work is properly cited*):

Factors	***OR***	***95% CI***	***p-value***
Age (control, 45) (years)			
55	**1.436**	**1.085–1.900**	**0.011**
65	**2.032**	**1.455–2.839**	**0.000**

75	**1.672**	**1.031–2.714**	**0.037**
Gender (control, male)	**0.765**	**0.595–0.986**	**0.038**
Residence (control, urban)	0.851	0.609–1.190	0.345
Education level (control, illiteracy)			
Primary school	0.911	0.685–1.213	0.525
Junior high school	0.745	0.543–1.022	0.068
Senior high school	0.701	0.442–1.112	0.131
University or higher	0.760	0.292–1.982	0.575
Income (control, lowest)			
Lower	1.264	0.912–1.750	0.159
Medium	0.926	0.660–1.299	0.656
Higher	0.759	0.530–1.086	0.131
Highest	1.039	0.709–1.524	0.844
Working status (control, working)			
Retirement	0.516	0.260–1.025	0.059
Unemployment	0.667	0.334–1.334	0.252
Medical insurance (control, no)			
	0.771	0.446–1.331	0.351
BMI (control, normal)			
Underweight	0.960	0.563–1.638	0.882
Overweight	**2.473**	**1.462–4.183**	**0.001**
Diabetes (control, no)			
	3.159	**2.129–4.687**	**0.000**
Exercise scoring (control, low intensity)			
Medium	0.775	0.568–1.057	0.107
High intensity	0.810	0.634–1.037	0.094
Smoking (control, no)	0.810	0.579–1.133	0.218
Depression (control, no)	1.231	0.971–1.561	0.086
Social support (control, lowest)			
Lower	**0.618**	**0.449–0.851**	**0.003**
Medium	**0.625**	**0.417–0.936**	**0.023**
Higher	0.733	0.516–1.041	0.082
Highest	**0.608**	**0.432–0.856**	**0.004**
Social participation (control, no)			
	1.191	0.955–1.486	0.120

The trend in hypertension and associated socioeconomic factors was assessed by Veisani and team in 690 adults in Ilam province in the *Journal of cardiovascular and thoracic research*. Analyses revealed a higher predominance of hypertension in lower socioeconomic groups. Further, the odds of hypertension were higher for older age (more than 66 years old vs. 15-24 years old), being male, living in urban settings vs. rural ones, being unemployed and illiteracy as shown in table 8 below that documents the results of analyzing for socioeconomic factors and hypertension in 690 participants (Veisani et al, 2019: Copyright © 2019 The Author(s): *This is an open access article distributed under the terms of the Creative Commons Attribution License (http://creativecommons.org/licenses/by/4.0/), which permits unrestricted use, distribution, and reproduction in any medium, provided the original work is properly cited*):

Key: OR; odds ratio, CI; confidence intervals

Independent variables	**Adjusted OR**	**95% CI**	***P*-value**
Age group (y)			
15-24	Reference	1	
25-44	1.36	1.21-1.60	<0.001
45-64	1.59	1.11-2.16	<0.001
≥65	1.89	1.19-2.87	<0.001
Sex			
Female	Reference	1	
Male	1.21	1.01-3.36	0.012
Residence			
Rural	Reference	1	
Urban	1.30	1.05-3.59	<0.001
Job			
Unemployment	Reference	1	
Employed	0.66	0.47-0.87	0.004
Educational level			
Illiterate	Reference	1	
Primary school	0.89	0.73-1.13	0.093
Guidance/high school	0.82	0.63-0.98	0.048
Diploma	0.73	0.52-0.91	0.001

University	0.72	0.68-0.97	0.001
Marital statues			
Marriage	Reference	1	
Single	0.91	0.63-0.84	<0.001
Economic status			
Low	Reference	1	
Middle	0.86	0.67-1.37	0.291
High	0.83	0.58-0.96	0.006

Finally, job (15%), education (18%), and economic status (21%) emerged as the causes of inequality. Figure 3 below documents the factors associated with hypertension as analyzed by research Veisani et al, 2019: Copyright © 2019 The Author(s): *This is an open access article distributed under the terms of the Creative Commons Attribution License (http:// creativecommons.org/licenses/by/4.0/), which permits unrestricted use, distribution, and reproduction in any medium, provided the original work is properly cited*):

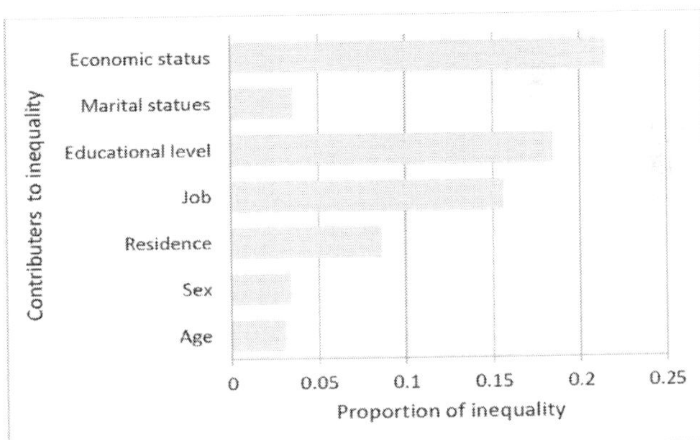

"For as long as our people are held hostage by controllable socio-economic forces, we cannot afford to be indifferent to the ravages of poverty in all its dimensions and ramifications"

-Ibrahim Babangida

1.5 VARIATIONS

Analysis of Clinical practice guidelines (CPGs) for hypertension between 2010 and 2020 in *BMC Medicine* by Philip and team (2021) across low-income countries (LICs), lower-middle-income countries (lower-MICs), upper-middle-income countries (upper-MICs) and high-income countries (HICs) on the basis of the World bank rating revealed that a target BP of <140/90 mmHg was recommended by 85% of the guidelines while ≥140/90 mmHg in the clinic was the definition of hypertension in 96% of these guidelines. When looking at the treatment options, a broad range of anti-hypertensive medication was offered in HICs while diuretics represented 63% of the recommendations in LICs.

In a nutshell, the authors reported major uniformity across hypertension definition and staging. However, when it came to treatment recommendations a stark difference emerged. Higher-income countries were offered dual therapy while lower-income countries saw a lower choice in medication with cheaper drugs being offered.

Table 9 below displays ***TOP:*** variations in the CPGs for the target BP for the elderly (Philip et al, 2021: *Open Access: This article is licensed under a Creative Commons Attribution 4.0 International License, which permits use, sharing, adaptation, distribution and reproduction in any medium or format, as long as you give appropriate credit to the original author(s) and the source, provide a link to the Creative Commons licence, and indicate if changes were made. The images or other third party material in this article are included in the article's Creative Commons licence, unless indicated otherwise in a credit line to the material. If material is not included in the article's Creative Commons licence and your intended use is not permitted by statutory regulation or exceeds the permitted use, you will need to obtain permission directly from the copyright holder. To view a copy of this licence, visit http://creativecommons.org/licenses/by/4.0/. The Creative Commons Public Domain Dedication waiver (http://creativecommons.org/publicdomain/zero/1.0/) applies to the data made available in this article, unless otherwise stated in a credit line to the data*):

Key: IGH Indian Guidelines for Hypertension, *JSH* Japanese Society of Hypertension, *ESC/ESH* European Society of Cardiology and Hypertension

Guideline	**Recommended target BP in the elderly**
Poland 2017 [58] Somalia 2015 [21]	SBP 140–150 mmHg
Thailand 2015 [39]	Age 60–80, 140–150/90 mmHg Age≥80, < 150/90 mmHg
China 2019 [44]	Age 65–80, < 140/90 mmHg Age≥80, < 150/90 mmHg

IGH India 2013 [27]	Age 55–79, < 140/90 mmHg Age > 80, SBP 140–140 mmHg
ISH 2020 [65] JSH/Japan 2019 [54]	< 140/90 mmHg
Korea 2018 [52]	SBP < 140 mmHg
ESC/ESH 2018 [56]	SBP 130–139 mmHg DBP 70–79 mmHg
Taiwan 2017 [60]	SBP < 140/90
ACC/AHA 2017 [55]	Age ≥ 65, SBP < 130 mmHg

BOTTOM: variations across treatment recommendations based on income (Philip et al, 2021: *Open Access: This article is licensed under a Creative Commons Attribution 4.0 International License, which permits use, sharing, adaptation, distribution and reproduction in any medium or format, as long as you give appropriate credit to the original author(s) and the source, provide a link to the Creative Commons licence, and indicate if changes were made. The images or other third party material in this article are included in the article's Creative Commons licence, unless indicated otherwise in a credit line to the material. If material is not included in the article's Creative Commons licence and your intended use is not permitted by statutory regulation or exceeds the permitted use, you will need to obtain permission directly from the copyright holder. To view a copy of this licence, visit http://creativecommons.org/licenses/by/4.0/. The Creative Commons Public Domain Dedication waiver (http://creativecommons.org/publicdomain/zero/1.0/) applies to the data made available in this article, unless otherwise stated in a credit line to the data*):

Income setting	**First-step therapy**	**Second-step therapy**	**Third-step therapy**
All (n = 48)	Any one of: • ACEi/ARB • CCB • Diuretic (33% of CPGs)	ACEi + CCB/diuretic (19% of CPGs)	ACEi/ARB + CCB + Diuretic (59% of CPGs)
Low-income countries (n = 8)	Diuretic (63% of CPGs)	No distinct majority (6 different combinations)	ACEi/ARB + CCB + Diuretic (60% of CPGs)

Lower-middle income countries (n=10)	CCB/diuretic (30% of CPGs)	CCB/diuretic + ACEi/ARB (30% of CPGs)	ACEi/ARB + CCB + Diuretic (83% of CPGs)
Upper-middle income countries (n=11)	Any one of: • ACEi/ARB • CCB • Diuretic • Beta-blocker (45% of CPGs)	No distinct majority (7 different combinations)	ACEi/ARB + CCB + Diuretic (50% of CPGs)
High-income countries (n=18*) *Including ISH 2020 optimal	Any one of: • ACEi/ARB • CCB • Diuretic (47% of CPGs)	No distinct majority (8 different combinations)	ACEi/ARB + CCB + Diuretic (50% of CPGs)

Research is warranted as to exploring the causes for this variation and the formulation of optimal treatment to quell these differences.

1.6 A WALK THROUGH HISTORY

A timeline of notable studies that looked at addressing hypertension was presented by Saklayen and Deshpande in *Frontiers in Cardiovascular Research* (2016) as presented below in ***TOP***: figurative approach (© 2016 Saklayen and Deshpande. *This is an open-access article distributed under the terms of the* ***Creative Commons Attribution License (CC BY)****. The use, distribution or reproduction in other forums is permitted, provided the original author(s) or licensor are credited and that the original publication in this journal is cited, in accordance with accepted academic practice. No use, distribution or reproduction is permitted which does not comply with these terms*):

Key: VACoop, Veterans Affairs Co-operative studies; HDFP, hypertension detection and follow-up trial; MRFIT, multiple risk factor intervention trial; MRC, Medical Research Council (UK); EWHPE, European Working Party High Blood pressure in the Elderly; SHEP, systolic hypertension in the elderly program; TOMHS, treatment of mild hypertension study; DASH, dietary approaches to stop hypertension; HOT, hypertension optimal treatment; UKPDS, United Kingdom Prospective Diabetes Study; AASK, African-American Study of Kidney Disease; ALLHAT, antihypertensive and lipid lowering treatment to prevent heart attack trial; ANBP, Australian National Blood Pressure Study; ASCOT-BPLA, Anglo-Scandinavian

Cardiac Outcomes Trial-Blood Pressure Lowering Arm; CAFÉ, Conduit Artery Function Evaluation; HYVET, hypertension in the very elderly trial; ACCOMPLISH, avoiding cardiovascular events through combination therapy in patients living with systolic hypertension; ACCORD, Action to Control Cardiovascular Risk in Diabetes; SPRINT, systolic blood pressure intervention trial

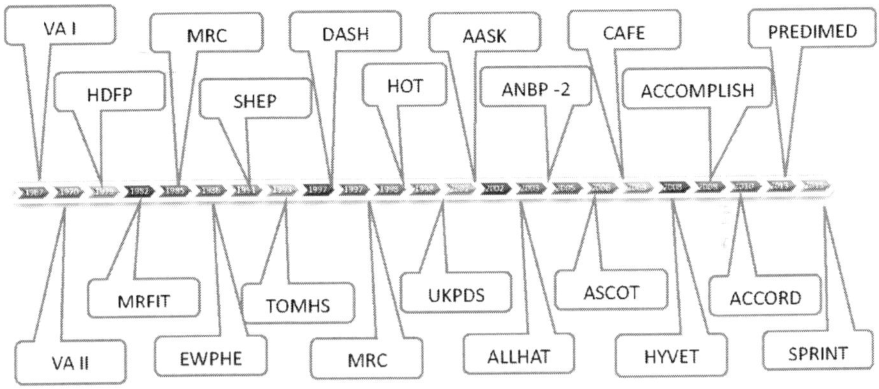

BOTTOM: Salient features of the approaches

Study	Year	Primary question/issues	Conclusion of the study/impact
VA-1st	1967	Is severe hypertension (dias) 115–129 treatable	Yes, less stroke/CHF
VA-2nd	1970	Same question for moderate BP (90–115)	Treated group less stroke/CHF
HDFP	1979	Goal-oriented BP therapy better than usual therapy?	Yes. Targeting BP goal of dias 90 reduced CVA by 36% more
MRFIT	1982	Lowering BP and lipid and stopping smoking may reduce CHD mortality	No difference in CHD mortality 17.9 vs. 19.3% (per 1000)
MRC	1985	Hypertension treatment in younger patients (35–64) is beneficial also?	Yes. Total CV events 286 in treated group vs. 352 in control ($p<0.05$)
EWHPE	1986	Hypertension treatment in exclusively older people (60) beneficial?	Yes. Mortality reduction 26% decrease in CV mortality 43%

SHEP	1991	Is treatment of systolic hypertension beneficial	Treating isolated systolic hypertension over 160 prevented stroke (ARR 3%), MI, and all CVD
TOMHS	1993	Outcome of 5 different classes BP meds vs. placebo	BP lowering similar among all classes CV events and death reduced (ARR 2.2%)
DASH	1997	Does Mediterranean diet with or without salt restriction lowers BP?	Compared to western diet it lowers bp and salt restriction adds to the effect
MRC	1997	Salt reduction in older people Lowers BP?	Reducing salt intake to 2 g Na lowered BP 7.2/3.2 mmHg
HOT	1998	Lowering Dias BP to 85 or 80 beneficial compared to standard 90 goal	No significant benefit in whole study but small benefit in diabetic
UKPDS	1998	Multiple studies 2 involved BP Tight BP control and agents (captopril vs. atenolol)	Group target <150/85 had 32, 44, and 34% less death, stroke, and retinopathy, respectively. No difference in ACEI group vs. BB
AASK	2002	To reduce progression of CKD BP goal mean 92 better than 105 ACEI, BB, or CCB better as drug?	No difference in mean BP goal of 92 vs. 105. ACEI use protected progression of CKD better than CCB
ALLHAT	2002	Compared to old thiazide (CTDN) new class of BP drugs CCB, ACEI, or AB has better outcome? AB gr closed for high incidence of CHF	No difference in MI, mortality, or CKD progression among 3 classes. CTDN vs. CCB for CHF RR 1.38. CTDN vs. ACEI for stroke and CHF RR 1.15 and 1.19
ANBP2	2003	ACEI vs. thiazide (HCTZ) for CV outcomes in Australian	In this study unlike ALLHAT, ACE was better all CV events RR was 0.88

ASCOT	2005	CCB and ACE inhibitor compared to BB and thiazide for BP control	CCB and ace inhibitor combination group showed better CV outcomes
CAFÉ	2006	Why Betablocker for BP does not prevent stroke	Betablocker lowers peripheral BP but not central (aortic) BP
HYVET	2008	Should we treat elderly (>80) hypertensive (sys>160)	Yes. Treated group had 30% less stroke and 64% less CHF, 21% less death
AC-SH	2008	Combination of ACEI+CCB better than ACEI+thiazide (HCTZ)	ACEI+CCB group had 2.2% ARR of composite CV events and death
ACCRD	2010	In diabetics Goal BP sys<120 better than 140?	No significant difference in mortality, total CV events, or renal protection
SPRINT	2015	Same as ACCORD but in non-diabetic	27% improved all-cause mortality and 25% improvement in primary CV outcomes

1.7 BARRIERS

A community-based, cross-sectional study at Kathmandu, Nepal explored resistance to hypertension treatment between January and July 2015. Devkota, Dhungana, Pandey, Bista, Panthi, Thakur and Gajurel (2016) reported that 191 of 587 participants had hypertension of which 118 were aware they had the condition (more than half were without formal education or did primary education). The consumption of anti-hypertensive medication displayed variation across income, age among other factors. The major barriers to taking treatment were the uncertainty of side effects and the requirement of lengthy treatment as displayed in table 10 below: ***TOP***: Control of hypertension and linked aspects © 2016 Devkota, Dhungana, Pandey, Bista, Panthi, Thakur and Gajurel. *This is an open-access article distributed under the terms of the* ***Creative Commons Attribution License (CC BY)****. The use, distribution or reproduction in other forums is permitted, provided the original author(s) or licensor are credited and that the original publication* *in this journal is cited, in accordance with accepted academic practice. No use, distribution or reproduction is permitted which does not comply with these terms*):

Characteristics	Categories	N_1	Treatment of hypertension among aware hypertensives n_1 (%)	*P* value	N_2	Control of hypertension among treated hypertensives n_2 (%)	*P* value
Total		118	93 (78.8)		93	46 (49.6)	
Gender	Male	54	42 (77.8)	0.48	42	20 (47.6)	0.77
	Female	64	51 (79.7)		51	26 (51.0)	
Age groups	<40 years	13	7 (58.8)	<0.01	7	5 (71.4)	0.48
	40–59 years	70	53 (75.7)		53	12 (12.9)	
	≥60 years	35	33 (94.3)		33	16 (48.5)	
Education	No formal education	48	36 (75.0)	0.76	36	20 (55.6)	0.08
	Primary and lower	19	15 (78.9)		15	4 (26.7)	
	Secondary	35	28 (80.0)		28	12 (42.9)	
	Bachelor and higher	16	14 (87.5)		14	10 (71.4)	
Ethnicity	Brahman	39	36 (92.3)	<0.01	36	19 (52.8)	0.65
	Chhetri	34	27 (79.4)		27	15 (56.6)	
	Newar	26	20 (76.9)		20	8 (40.0)	
	Others (Dalit, Janjati)	19	10 (52.6)		10	4 (40.0)	
Occupation	Government job	17	12 (70.6)	<0.01	12	8 (66.7)	0.54
	Self employed	19	10 (52.6)		10	5 (50.0)	
	Homemaker	61	51 (83.6)		51	25 (49.0)	
	Others	21	20 (95.2)		20	8 (40.0)	
Poverty	Above poverty line	62	54 (87.1)	0.04	39	15 (38.5)	0.07
	Below poverty line	56	39 (69.6)		54	31 (57.4)	
Smoking	Yes	28	17 (60.7)	0.01	17	8 (47.1)	0.83
	No	90	76 (84.4)		76	38 (50.0)	

Alcohol consumption	Yes	32	21 (65.6)	0.03	21	9 (42.9)	0.49
	No	86	72 (83.7)		72	37 (51.4)	
Fruit and vegetables intake	Sufficient	15	13 (86.7)	0.42	13	3 (23.1)	0.08
	Non-sufficient	103	80 (77.7)		80	43 (53.8)	
Physical activity	Sufficient	85	64 (75.3)	0.13	64	34 (53.1)	0.30
	Non-sufficient	33	29 (87.9)		29	12 (41.4)	
BMI	≥25 kg/m²	74	56 (75.7)	0.28	56	29 (51.8)	0.58
	<25 kg/m²	44	37 (84.1)		37	17 (45.9)	
Diabetes	Yes	23	20 (87.0)	0.29	20	11 (55.0)	0.58
	No	95	73 (76.8)		73	35 (47.9)	

N_1, aware hypertensive participants; N_2, aware hypertensive participants who were under treatment; n_1, proportions of participants who were under treatment; n_2, proportions of participants having controlled blood pressure; primary and lower education, grade 1–5; secondary, grade 6–12; job, governmental or non-governmental employment; self-employed, working for oneself as a freelancer or the owner of a business or working in own farm; Homemaker, a person, especially a housewife, who manages a home; other occupation, volunteer or student or unemployed or retired; Below poverty line, income less than 40,933 Nepalese rupees per annum; sufficient fruit and vegetable consumption, intake of ≥400 g of fruit and vegetable per day; Sufficient physical activity, ≥600 METs of moderate and vigorous activity in a week.

BOTTOM: Resistance to the control of hypertension and linked aspects

Characteristics	**Categories**	**Control of hypertension**		**Total**	***P* value**
		Yes (%)	**No (%)**		
Antihypertensive medicine	Mono-therapy	35 (44.9)	43 (55.1)	78	0.04
	Combination or Two and more medicines	11 (73.3)	4 (26.7)	15	
Access to health facility	<20 min	16 (51.6)	15 (48.4)	31	0.70
	≥20 min	27 (47.4)	30 (52.6)	57	
Waiting time in health facility	<20 min	22 (66.7)	11 (33.3)	33	0.03
	≥20 min	22 (42.3)	30 (57.7)	52	
Adherence to medication	High	30 (78.9)	8 (21.1)	38	<0.01
	Medium	16 (30.8)	36 (69.2)	52	
	Low	0 (0)	3 (100.0)	3	
Types of health-care providers	Cardiologist	30 (55.6)	24 (44.4)	54	0.17
	Others	16 (41.0)	23 (59.0)	39	
Types of health facilities	Government	12 (44.4)	15 (55.6)	27	0.54
	Other	34 (51.5)	32 (48.5)	66	
Follow-up continuation	Yes	38 (60.3)	25 (39.7)	63	<0.01
	No	7 (24.1)	22 (75.9)	29	
Satisfaction with service provider	Yes	43 (52.4)	39 (47.6)	82	0.18
	No	3 (27.3)	8 (72.7)	11	

Health-care provider advice on lifestyle modification	Yes	29 (61.7)	18 (38.3)	47	0.03
	No	17 (37.0)	29 (63.0)	46	

In another 2019-published research, Sibomana, J.P., McNamara, R.L. & Walker, T.D presented the **barriers** to efficient control of hypertension in a **rural** setting (4 rural hospitals in Rwanda) across 112 patients of whom only 29% had reached the recommended control. Table 11 below is representative of key findings that can help formulate appropriate systems to transcend these barriers:

TOP: Adherence to medication and responses in the rural settings (Sibomana et al, 2019: *Open Access: This article is distributed under the terms of the Creative Commons Attribution 4.0 International License (http://creativecommons.org/licenses/by/4.0/), which permits unrestricted use, distribution, and reproduction in any medium, provided you give appropriate credit to the original author(s) and the source, provide a link to the Creative Commons license, and indicate if changes were made. The Creative Commons Public Domain Dedication waiver (http://creativecommons.org/publicdomain/zero/1.0/) applies to the data made available in this article, unless otherwise stated*):

Question	**Number of "No"**
Do you sometimes forget to take your medicine? (No=0)	16 (14)
People sometimes miss taking their medicines for reasons other than forgetting. Thinking over the past 2 weeks, were there any days when you did not take your medicine? (No=0)	9 (8.0)
Have you ever cut back or stopped taking your medicine without telling your doctor because you felt worse when you took it? (No=0)	7 (6.2)
When you travel or leave home, do you sometimes forget to bring along your medicine? (No=0)	6 (5.4)
Did you take all your medicines yesterday? (Yes=0)	106 (94.6)
When you feel like your symptoms are under control, do you sometimes stop taking your medicine? (No=0)	4 (3.6)
Taking medicine every day is a real inconvenience for some people. Do you ever feel hassled about sticking to your treatment plan? (No=0)	1 (0.9)
Never/rarely (0)	99 (88.4)

Once in a while (1)	10 (8.9)
Sometimes (2)	2 (1.8)
Usually (3)	1 (0.9)
Always (4)	0
High (Total score=0)	86 (76.8)
Low to moderate (Total score≥1)	26 (23.2)

BOTTOM: Barriers to achieving BP control; N=30 clinicians except at * where N=29 (Sibomana et al, 2019: *Open Access: This article is distributed under the terms of the Creative Commons Attribution 4.0 International License (http://creativecommons.org/licenses/by/4.0/), which permits unrestricted use, distribution, and reproduction in any medium, provided you give appropriate credit to the original author(s) and the source, provide a link to the Creative Commons license, and indicate if changes were made. The Creative Commons Public Domain Dedication waiver (http://creativecommons.org/publicdomain/zero/1.0/) applies to the data made available in this article, unless otherwise stated*):

Domain	Perceived Barriers	*N* (%)
Patient	Poor adherence to medication therapy	19 (63.3)
	Lack of knowledge about hypertension	23 (76.7)
	Inability to engage in lifestyle changes	23 (76.7)
	Health beliefs	7 (24.1)*
	Medication side effects	6 (20.0)
Clinician	Nonadherence to treatment guidelines	15 (50.0)
	Failure to emphasize lifestyle modifications	10 (34.5)
	Clinical inertia	12 (41.4)
Logistic	Unavailability of prescribed medication in the hospital pharmacy	23 (76.7)
	Lack of access to care	17 (56.7)
	High cost of medications	19 (63.3)
	Absence of clinical decision support systems	13 (43.3)
	High copayments	9 (30.0)

A 2019-published article described as "*the first large-scale population-based study that examines health system performance in management of hypertension in India*" examined the trend in hypertension employing 731,864 participants from the 2015–2016 National Family Health Survey (NFHS-4) data. The total prevalence of hypertension was 18.1% with the prevalence in men at 19.0% vs. 17.2% in ladies. Controlled BP was reported by 7.9% with hypertension care being availed more in urban areas and by women. Further, it was found that 76% of hypertensive individuals had their BP measured with 45% having an awareness that they have hypertension and 13% were on medication with BP control seen only in 8%.

Figure 4 below is illustrative of the hypertension care across the NFHS-4 data scanned while table 12 below represents the predictors of hypertension control (© 2019 Prenissl et al. *This is an open access article distributed under the terms of the Creative Commons Attribution License, which permits unrestricted use, distribution, and reproduction in any medium, provided the original author and source are credited*):

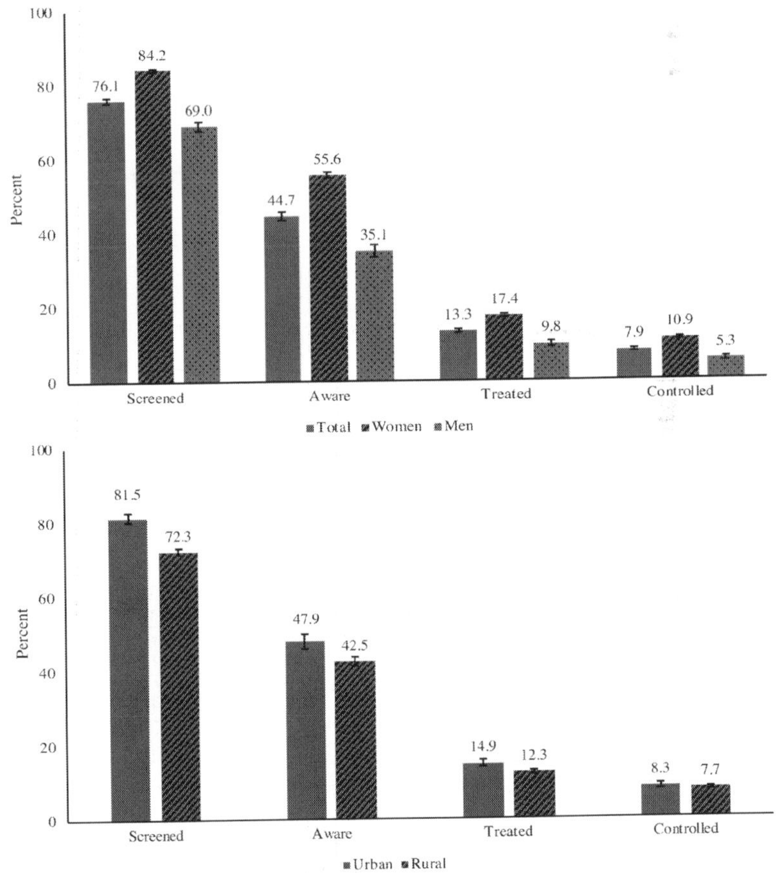

Characteristic	Treated hypertension				Controlled hypertension			
	Rural		Urban		Rural		Urban	
	RR (95% CI)	*P* value	RR (95% CI)	*P* value	RR (95% CI)	*P* value	RR (95% CI)	*P* value
Age group								
15–19 years	1.00 (reference)		1.00 (reference)		1.00 (reference)		1.00 (reference)	
20–24 years	0.90 (0.81–0.99)	0.032	0.98 (0.82–1.16)	0.784	0.88 (0.79–0.98)	0.017	0.98 (0.81–1.18)	0.809
25–29 years	0.94 (0.85–1.03)	0.193	0.88 (0.74–1.04)	0.133	0.89 (0.80–0.99)	0.036	0.82 (0.67–0.99)	0.035
30–34 years	0.96 (0.87–1.07)	0.473	1.01 (0.85–1.18)	0.952	0.84 (0.75–0.94)	0.002	0.81 (0.67–0.98)	0.028
35–40 years	1.06 (0.96–1.17)	0.267	1.16 (0.99–1.36)	0.067	0.78 (0.70–0.88)	<0.001	0.78 (0.64–0.94)	0.008
40–44 years	1.19 (1.08–1.31)	<0.001	1.53 (1.31–1.79)	<0.001	0.77 (0.69–0.86)	<0.001	0.94 (0.78–1.13)	0.518
45–49 years	1.38 (1.25–1.52)	<0.001	1.94 (1.65–2.26)	<0.001	0.79 (0.71–0.89)	<0.001	1.15 (0.96–1.38)	0.141
Education								
Primary school unfinished	1.00 (reference)		1.00 (reference)		1.00 (reference)		1.00 (reference)	
Primary school finished	1.01 (0.95–1.08)	0.652	1.05 (0.97–1.15)	0.243	1.02 (0.93–1.11)	0.726	1.07 (0.95–1.21)	0.264
Secondary school unfinished	1.03 (0.99–1.08)	0.151	0.98 (0.92–1.04)	0.503	1.06 (1.00–1.12)	0.064	0.98 (0.90–1.07)	0.643
Secondary school or above	1.00 (0.94–1.06)	0.974	0.96 (0.90–1.03)	0.293	1.06 (0.98–1.14)	0.158	0.97 (0.88–1.07)	0.583
Household wealth quintile								
Q1 (poorest)	1.00 (reference)		1.00 (reference)		1.00 (reference)		1.00 (reference)	
Q2	1.26 (1.17–1.36)	<0.001	1.11 (1.03–1.20)	0.009	1.21 (1.11–1.32)	<0.001	1.15 (1.03–1.28)	0.011

Q3	1.34 (1.24–1.44)	<0.001	1.17 (1.08–1.27)	<0.001	1.30 (1.19–1.42)	<0.001	1.15 (1.03–1.29)	0.012
Q4	1.44 (1.34–1.55)	<0.001	1.27 (1.17–1.38)	<0.001	1.38 (1.25–1.51)	<0.001	1.37 (1.22–1.54)	<0.001
Q5 (richest)	1.53 (1.42–1.66)	<0.001	1.34 (1.22–1.47)	<0.001	1.44 (1.30–1.59)	<0.001	1.40 (1.24–1.59)	<0.001
BMI								
<18.5 kg/m2	1.10 (1.04–1.16)	<0.001	1.09 (0.98–1.21)	0.128	1.19 (1.12–1.26)	<0.001	1.15 (1.01–1.30)	0.030
18.5–22.9 kg/m2	1.00 (reference)		1.00 (reference)		1.00 (reference)		1.00 (reference)	
23.0–24.9 kg/m2	1.02 (0.97–1.07)	0.448	1.01 (0.98–1.21)	0.767	0.86 (0.81–0.92)	<0.001	0.91 (0.83–1.00)	0.049
25.0–27.4 kg/m2	1.07 (1.01–1.12)	0.012	1.11 (1.04–1.19)	0.002	0.82 (0.77–0.88)	<0.001	0.90 (0.83–0.99)	0.030
27.5–29.9 kg/m2	1.17 (1.10–1.24)	<0.001	1.28 (1.19–1.37)	<0.001	0.86 (0.79–0.94)	0.001	0.98 (0.89–1.08)	0.658
≥30.0 kg/m2	1.46 (1.38–1.54)	<0.001	1.61 (1.51–1.71)	<0.001	1.04 (0.95–1.13)	0.405	1.23 (1.13–1.35)	<0.001
Tobacco consumption								
Current smoker	0.93 (0.86–1.01)	0.105	1.02 (0.91–1.14)	0.738	1.01 (0.90–1.13)	0.878	0.98 (0.82–1.15)	0.771
Uses smokeless tobacco	0.98 (0.92–1.04)	0.461	0.99 (0.92–1.07)	0.753	0.93 (0.86–1.01)	0.069	1.01 (0.90–1.13)	0.910
Currently married	1.11 (1.05–1.17)	<0.001	1.09 (1.02–1.16)	0.010	1.12 (1.04–1.20)	0.001	1.13 (1.03–1.24)	0.008
Female	1.78 (1.66–1.91)	<0.001	1.64 (1.51–1.79)	<0.001	2.01 (1.83–2.21)	<0.001	1.95 (1.72–2.20)	<0.001

These Poisson regressions contained all variables listed in the table (age group, household wealth quintile, education, BMI, tobacco consumption, marital status, and sex) and a binary indicator for each district (district-level fixed effects) as predictor variables. Standard errors were adjusted for clustering at the level of the primary sampling unit. Regression results including an interaction term between marital status and sex, using sampling weights (to account for the higher number of sampled women), and running analyses separately for women and men can be found in Tables X–AA in S1 Supplemental Materials. The sample for each regression was all individuals with hypertension (i.e., the sample was not restricted to those who had reached the preceding cascade step).

BMI, body mass index; RR, risk ratio.

Such lack of awareness and lack of controlled BP as reported by research raises an important question as to the availability and knowledge of diseases. Further research and plans are warranted that are aimed at reducing such gaps.

1.8 RECOMMENDATION GUIDELINES

In order to facilitate timely and precise diagnosis of hypertension and monitoring, guidelines from several countries have been compiled by Schwartz and McManus (2015). These have been presented below in table 13 (Schwartz and McManus, 2015: *Open Access: This article is distributed under the terms of the Creative Commons Attribution 4.0 International License (http://creativecommons.org/licenses/by/4.0/), which permits unrestricted use, distribution, and reproduction in any medium, provided you give appropriate credit to the original author(s) and the source, provide a link to the Creative Commons license, and indicate if changes were made. The Creative Commons Public Domain Dedication waiver (http://creativecommons.org/publicdomain/zero/1.0/) applies to the data made available in this article, unless otherwise stated*):

Key: OBPM office blood pressure measurement, *ABPM* ambulatory blood pressure measurement, *HBPM* home blood pressure measurement, *CKD* chronic kidney disease

Organisation	Year	Diagnostic threshold	Treatment threshold
UK National Institute for Health and Care Excellence (NICE)	2011	OBPM diagnostic threshold: ≥140/90 mmHg (ABPM/HBPM diagnostic threshold: ≥135/85 mmHg)	All patients under 80 yrs: OBPM< 140/90 mmHg (ABPM/HBPM: <135/85 mmHg)
		OBPM Stage 2 Hypertension ≥160/100 mmHg (ABPM/HBPM: ≥150/95 mmHg)	Diabetes : OBPM <140/80 mmHg (or <130/80 mmHg if complications present)
			Older ≥80 yrs: OBPM <150/90 mmHg (ABPM/HBPM <145/85 mmHg)

European Society of Hypertension (ESH)	2013	OBPM: ≥140 and/or ≥90	All patients under 80 yrs OBPM <140/90 mmHg
		(ABPM Daytime (or awake): ≥135 and/or ≥85 mmHg, ABPM Night-time (or asleep): ≥120 and/or ≥70 mmHg, ABPM 24-h: ≥130 and/or ≥80 mmHg, HBPM: ≥135 and/or ≥85 mmHg)	Diabetes: OBPM <140/85 mmHg
			High Risk Patients: OBPM <130/80 mmHg
			Older ≥80 yrs: OBPM <150/90 mmHg
Joint National Committee on Prevention, Detection, Evaluation and Treatment of Blood Pressure (JNC 7)	2004	Stage 1 hypertension diagnosis should be confirmed within 2 months after initial elevated OBPM ≥140/90 mmHg	Non-diabetic patients: OBPM <140/90 mmHg
			Diabetic/CKD patients: OBPM <130/80 mmHg
		Stage 2 hypertension should be confirmed within 1 m	
		≥180/110 mm Hg evaluate and treat immediately	
		(ABPM Daytime (or awake): ≥135/85 mmHg, ABPM Night-time (or asleep): ≥120/75 mmHg)	
		Routine blood pressure measurements should be taken	
		– at least once every 2 years for adults with <120/80 mmHg	
		– every year for adults with 120-139/80-89 mmHg	
Joint National Committee on Prevention, Detection, Evaluation and Treatment of Blood Pressure (JNC 8)	2014	• Guidelines did not address diagnostic thresholds of hypertension.	Age <60 yrs: OBPM <140/90 mmHg
			Diabetes No CKD: OBPM <140/90 mmHg
		• The supplementary material recommends averaging 2–3 measurements at each visit to establish a diagnosis of hypertension.	CKD present with or without diabetes: OBPM <140/90 mmHg
			Older ≥60 years: OBPM <150/90 mmHg
		• Thresholds for pharmacological treatment were defined.	
		• HBPM and ABPM were not included.	

Canadian Hypertension Education Program (CHEP)	2015	ABPM Daytime (or awake): ≥135/85 mmHg	All ages <80 yrs: OBPM <140/90 mmHg
		ABPM 24-h: ≥130/80 mmHg	Diabetes: OBPM <130/80 mmHg
		HBPM diagnostic threshold: ≥135/85 mmHg)	Older ≥80 yrs: OBPM <150 mmHg
		OBPM diagnostic threshold:	
		≥140/90 mmHg averaged across two visits;	
		≥160/110 mmHg averaged across three visits;	
		or if ≥140/90 mmHg averaged across five visits	

1.9 SMARTPHONES AND HYPERTENSION MONITORING

Given the usage of smartphones today, such mobile-based approaches to quantify blood pressure were reviewed by Lee and team of the Korean Society of Hypertension was published in *Clinical Hypertension*. A search conducted between January 2010 and July 2020 revealed the following studies, the details of which are summarized by the authors below (Lee et al, 2021: *Open Access: This article is licensed under a Creative Commons Attribution 4.0 International License, which permits use, sharing, adaptation, distribution and reproduction in any medium or format, as long as you give appropriate credit to the original author(s) and the source, provide a link to the Creative Commons licence, and indicate if changes were made. The images or other third party material in this article are included in the article's Creative Commons licence, unless indicated otherwise in a credit line to the material. If material is not included in the article's Creative Commons licence and your intended use is not permitted by statutory regulation or exceeds the permitted use, you will need to obtain permission directly from the copyright holder. To view a copy of this licence, visit http://creativecommons.org/licenses/by/4.0/. The Creative Commons Public Domain Dedication waiver (http://creativecommons.org/publicdomain/zero/1.0/) applies to the data made available in this article, unless otherwise stated in a credit line to the data*):

Year	Authors	Subjects	Type of subjects	Device	Comparison (Measuring location / device)	Mean±SD (mmHg)	
						SBP	DBP
2013	Chandrasekaran, et al. [12]	500	Healthy subjects	Nexus One	Brachial cuff / Mercury sphygmomanometer	±2.4473 (95% CI)	±1.7073 (95% CI)
2014	Visvanathan, et al. [13]	156	Variable subjects	Iphone4	Brachial cuff / ET-COMM HC-502	0.78 (F-score)	0.80 (F-score)
2016	Plante, et al. [14]	85	Variable subjects	iPhone 5 s, 6	Brachial cuff / Omron 907, 907XL	12.4±10.5	10.1±8.1
2016	Gaurav, et al. [15]	3000	Healthy subjects	galaxy note 5	Brachial cuff / Mercury sphygmomanometer	0.16±6.85	0.03±4.72
2016	Alexander, et al. [16]	100	Healthy subjects	iPhone 5 s	Brachial cuff / Welch Allyn Propaq CS Monitor	−0.6±12.8	+7.2±9.2
2016	Gao, et al. [17]	78	Healthy subjects	Android phone	Brachial cuff / A&D UA-767PBT	5.1±4.3	4.6±4.3
2018	Raichle, et al. [18]	32	Variable subjects	Iphone 4 s	Brachial cuff / Omron-HBP-1300	5.0±14.5	NA.
2018	Wang, et al. [19]	7	Variable subjects	Google pixel phone	Brachial cuff / Microlife BP3NA1-1x	N.A.	4.8±4.7
2018	Chandrasekaran, et al. [20]	20	Healthy subjects	iPhone X	Brachial cuff / Omron BP786	−4.0±11.4	−9.4±9.7
2018	Dey, et al. [21]	205	Variable subjects	Galaxy S6	Brachial cuff / Mercury sphygmomanometer	6.90±9.00	5.00±6.10
2018	Matsumura, et al. [22]	49	Variable subjects	iPhone 6 s	Brachial cuff / NISSEI, DS-S10	0.67±12.7	0.45±8.6
2019	Luo, et al. [23]	1328	Healthy subjects	iPhone 6+	Finger sensor / CNAP 500	0.39±7.3	−0.20±6.00

While the accuracy of these approaches used was reported to be satisfactory in order to augment awareness of hypertension the "other side" exists too given the still ongoing research and the possibility of incorrect measurement by users. However, the "dynamic" and "smart" measurement of hypertension in the future looks promising!

1.10 PATIENT AND PHYSICIAN PERSPECTIVES

Hypertensive ***patients*** (in the age range of 26–84 years; n=26) from two hypertension clinics at the University of Alberta (Edmonton, Midwestern Canada) were subjected to qualitative interviews in order to gauge their viewpoint (Jolles et al, 2013). This 2013-published research outlined interesting perspectives that can help understand their routines and beliefs that are summarized below:

Table 14 below outlines the **feedback produced verbatim of hypertensive patients** in the routine management of hypertension (Copyright © 2013 Emily P. Jolles et al. *This is an open access article distributed under the Creative Commons Attribution License, which permits unrestricted use, distribution, and reproduction in any medium, provided the original work is properly cited*):

Theme	Quotes
Routine	*P: I've never missed it. I get right panicky if I have to miss it. And I know, you know the doctors have said, if you miss one day it's not going to, you know, but I don't like to miss no. I don't miss.*
	I: What are you worried about if you miss?
	P: Actually I can't tell you what I worry about, I just think it's not probably a good thing. (female, on target)
Routine	*It really doesn't happen unless [—] um if schedules are all screwed up, if we're away somewhere or travelling yeah. Which doesn't happen more than you know, a couple times a year. (female, not on target)*
Monetary issues	*P: Well it's not cheap. And then you know, it's like taking the medication right? I can't even afford it right now. The healthier I stay, the better, cheaper it is for me. (male, not on target)*
Routine-support	*Yes sometimes wife tell me [—] [name] take all pills! Alright told [name] are you taking all together like this? (male, not on target)*
Routine-coping	*I'm a frequent traveler so I carry my pills. Take them, take them and yeah I'm [—] very rarely might I miss a day of taking them but most of the time I take them. (male, on target)*

Routine-coping	*But it's quite easy when you have them in your little containers. (female, not on target)*
Routine-coping	*Oh I for years was really bad. I have a dosette now and I faithfully, got it all numbered the days of the month. (male, not on target)*
Routine	*It's not going to be a problem. I've been doing it for years, So it's part of my ritual. It's just the way it is. You have high blood pressure and you take it you don't forget. (male, on target)*
Life-style-barriers	*Um [—] but of course really the only setback is to that is because of my lifestyle I am very busy, and you don't always have time to eat healthy. But that's, you know, you know I try and eat healthy and exercise as much as I can. I think that's the part that stresses me out a little bit is that [—] um you know, I know I should be losing weight and I know I should be doing it for my health, but I just have to get to I know it's important medically. Um [—] you know but with lifestyle it is really hard to do that. (patient 13, female, on target)*
Routine-lifestyle	*Yeah sometimes weekends aren't the best for me [laughs] And I'm a bit worried about that if I retire 'cause I think sometimes um [—] I eat a better diet on workdays because I have to make a lunch and take it. [—] I think I'm just on more of a routine. (female, not on target)*
Dietary adher-ence-spousal support	*P: And salt free diet, or salt reduced.*
	I: And how do you follow a salt free diet?
	P: My wife takes care of it. (male, on target)

Table 15 below outlines the **feedback** produced verbatim of hypertensive patients in their **awareness of hypertension** (Copyright © 2013 Emily P. Jolles et al. *This is an open access article distributed under the Creative Commons Attribution License, which permits unrestricted use, distribution, and reproduction in any medium, provided the original work is properly cited*):

Theme	Quotes
Knowl-edge	*Well it is definitely not a good thing for the body. I mean it can cause a lot of problems to kidneys and strokes and you know make your heart work faster and therefore weaken it. You know, it is great when you exercise and you get your heart going faster, but when your heart has to pump faster just to, kind of you know, deal with your everyday life I mean it is not, it is not very clever at all. It is not very beneficial to you to have high blood pressure. (female, on target*)
Knowl-edge and HCP	*P: Yeah I suppose I do not ask them because [—] I, I they will take the readings and they will vary. I assume that if, if they had concerns they would ask me, and I do not want to be a hypochondriac or be a [chuckles] asking rhetorical questions that do not really have an answer to them, so. And I sort of respect their time so [—] and, and they might seem like silly questions so.*
	I: Do you get that feeling from something that they did or is that more?
	P: No, just respect their time and know that [—] I just maybe think they are [—] dumb questions perhaps or taking up the doctor's time asking so. (male, on target)
Knowl-edge and HCP	*I: Do you have a good understanding about what high blood pressure is? So if your doctor was in the room right now would you have questions for him that you still do not really understand*?
	P: What causes high blood pressure? Right. Why do they put you on medication, things like that.
	I: Right.
	P: They should answer that and talk about it.
	P: The only person who will do that [—] is a pharmacist. They will explain it more than the doctors. 'Cause the doctors are in and out, they're so busy. (male, not on target)
Knowl-edge: BP medica-tion	*I: Do you have any idea of what it does for you? Or how it works or?*
	P: No. I do not. (male, not on target)
Knowl-edge: BP medica-tion	*I can't say as I do [know what each pill does], I know it's supposed to lower my blood pressure. So I know that the water pill is very helpful. Um other than that, I do not know technically what mine does. (female, on target)*

Side effects	*Um, no I mean [—] if the side effects, if the sexual thing was a proven side effect [—] I would have to make a decision because [—] I might go off it just to see if it changed much. (male, on target)*
Side effects	*Yeah. No the other ones I was on were nasty. Like within, within four hours of taking it, I would be head over the toilet bowl, sick, migraine [—] nasty. ··· but my hair was falling out and the pharmacist said, yeah the diuretic can do that. So I asked my doctor, please change. And he went through this whole mess of all these other ones that we were trying that just made me deathly ill. (female, not on target)*
Knowledge-lifestyle	*P: Yeah, well every time I go to the doctor [—] he always goes over it and you know tells me what I should be doing. And I guess, if I did exactly what he said, more than likely I would not have high blood pressure. (male, not on target)*
Knowledge-lifestyle	*P: No I mean cognitively I know that stop smoking and lose weight and get active is the best thing you can do for blood pressure. I am sure eating well and cutting out is probably another good thing too but I like salt so, I mean I know what I should be doing it's just that I am not doing it.*
	I: Is that what it is or what is getting in the way?
	P: Nothing, but sheer apathy.
	P: Or [—] laziness. I sit at work all day.
	P: Maybe something happening? You'd think that two hundred over whatever it was blood pressure would make me pick up and notice and start doing things but, I do not know, I really don't know.
	P: It is stupid, it is really stupid.
	I: Not to do anything or?
	P: Yes! I am sixty for God's sake. I mean life is not getting any shorter or any longer. Do I need to be taking care of this now. (female, on target)
Knowledge-lifestyle	*P: Yeah. I think, I do not know, maybe somebody would have to tell me, if you do not stop eating so much salt, you're going to be dead in five years.*
	I: Okay.
	P: And then I'd be like oh crap,
	I: So that would be your motivation?
	P: Yeah. (female, not on target)

Takuo Yoshida and team (2020) analyzed perspectives from PARADOX: Perspectives of patients And physicians RegArDing hypertensive management from an Online survey for eXcellence to comprehend the gaps in reaching blood pressure control despite therapeutic advances. The survey of 541 ***physicians*** revealed that lifestyle modifications and adherence to guidelines were followed more by specialists vs. non-specialist physicians. The burden on patients was more acknowledged by general practitioners vs. physicians in hospitals.

Figure 5 below shows the causes the physicians surveyed stated for optimal hypertension management: a) patient (the question here was *"To what extent do you think each of the following patient-derived and disease-related reasons are critical factors that prevent 100% of your patients from achieving the guideline-recommended target*?"). b) physician (the question here was "*To what extent do you think each of the following pharmacist- and physician-derived reasons are critical factors that prevent 100% of patients from achieving this target?*") : *BP blood pressure, DBP diastolic blood pressure, SBP systolic blood pressure.* $*P<0.05$; $**P<0.01$ (Yoshida et al, 2020: *Open Access: This article is licensed under a Creative Commons Attribution 4.0 International License, which permits use, sharing, adaptation, distribution and reproduction in any medium or format, as long as you give appropriate credit to the original author(s) and the source, provide a link to the Creative Commons license, and indicate if changes were made. The images or other third party material in this article are included in the article's Creative Commons license, unless indicated otherwise in a credit line to the material. If material is not included in the article's Creative Commons license and your intended use is not permitted by statutory regulation or exceeds the permitted use, you will need to obtain permission directly from the copyright holder. To view a copy of this license, visit http://creativecommons.org/licenses/by/4.0/*):

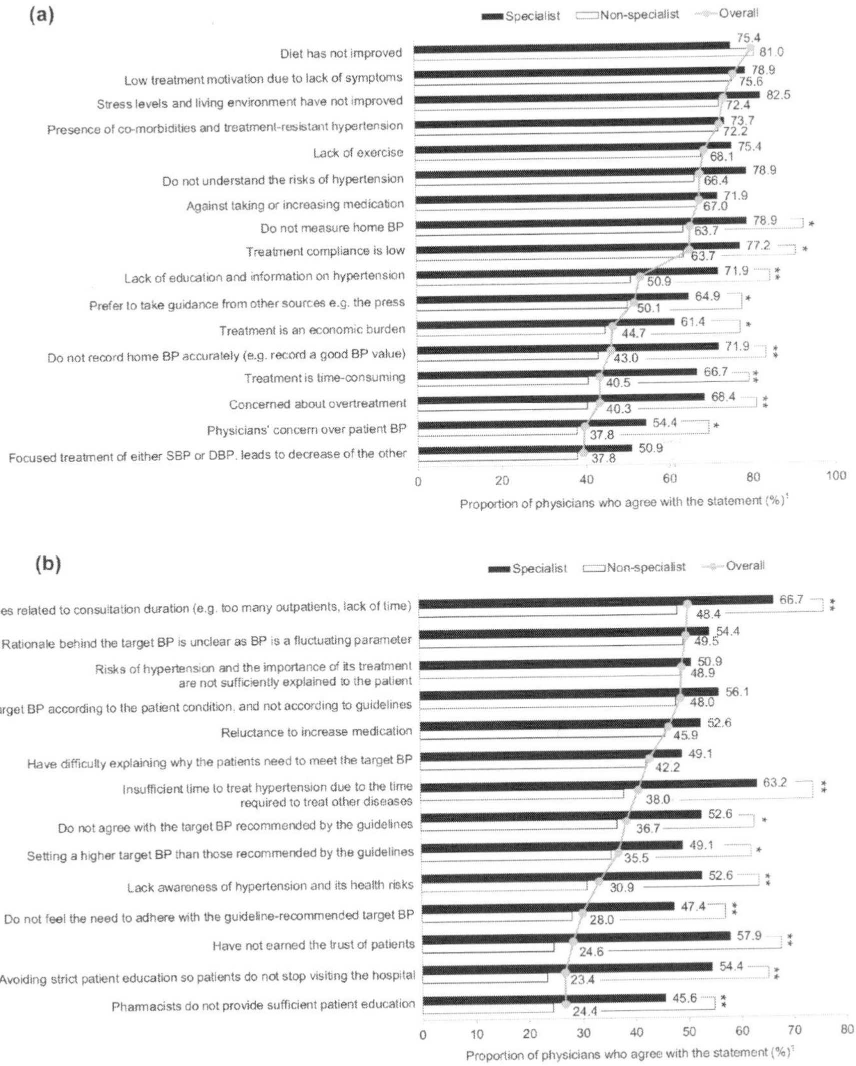

Figure 6 below shows the actions the physicians surveyed would look at in the future to help hypertension management: *$P<0.05$ (Yoshida et al, 2020: *Open Access: This article is licensed under a Creative Commons Attribution 4.0 International License, which permits use, sharing, adaptation, distribution and reproduction in any medium or format, as long as you give appropriate credit to the original author(s) and the source, provide a link to the Creative Commons license, and indicate if changes were made. The images or other third party material in this article are included in the article's Creative Commons license, unless indicated otherwise in a credit line to the material. If material is not included in the article's Creative*

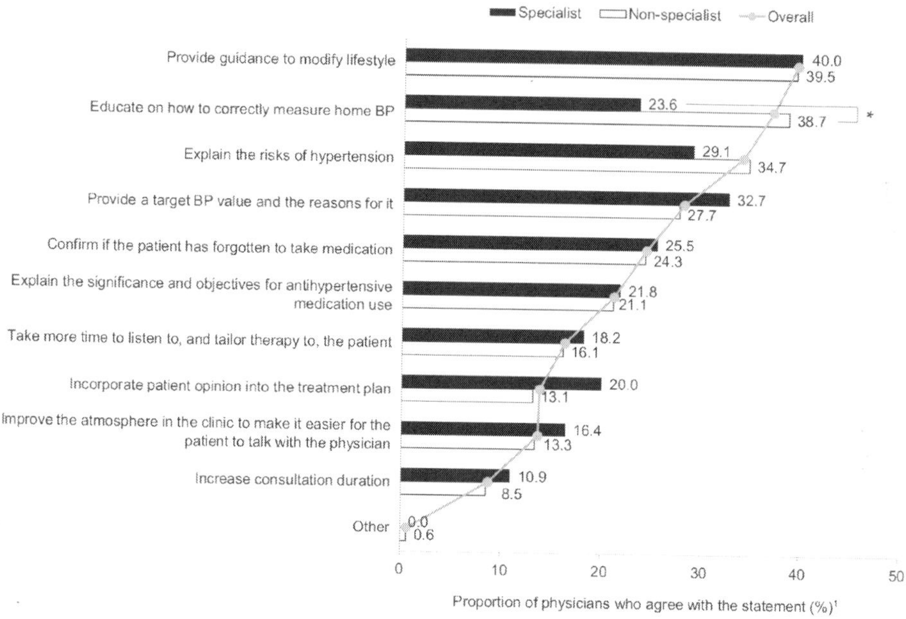

Combining perspectives from patients and physicians can help guide the formulation of optimal approaches to help achieve hypertension management.

"Let your hopes, not your hurts, shape your future."

—Robert H. Schuller

1.11 REFERENCES

1. Zeng, Z., Chen, J., Xiao, C. and Chen, W., 2020. A Global View on Prevalence of Hypertension and Human Develop Index. *Annals of Global Health*, 86(1), p.67. DOI: http://doi.org/10.5334/aogh.2591
2. Iqbal AM, Jamal SF. Essential Hypertension. [Updated 2020 Jul 10]. In: StatPearls [Internet]. Treasure Island (FL): StatPearls Publishing; 2021 Jan-. Available from: https://www.ncbi.nlm.nih.gov/books/NBK539859/
3. Majid Nour, Kemal Polat, "Automatic Classification of Hypertension Types Based on Personal Features by Machine Learning Algorithms", *Mathematical Problems in Engineering*, vol. 2020, Article ID 2742781, 13 pages, 2020. https://doi.org/10.1155/2020/2742781
4. Liu, J. Highlights of the 2018 Chinese hypertension guidelines. *Clin Hypertens* 26, 8 (2020). https://doi.org/10.1186/s40885-020-00141-3
5. Alwan H, Pruijm M, Ponte B, Ackermann D, Guessous I, Ehret G, et al. (2014) Epidemiology of Masked and White-Coat Hypertension: The Family-Based SKIPOGH Study. PLoS ONE 9(3): e92522. https://doi.org/10.1371/journal.pone.0092522
6. Xavier Trudel, Alain Milot, Chantal Brisson, "Persistence and Progression of Masked Hypertension: A 5-Year Prospective Study", International Journal of Hypertension, vol. 2013, Article ID 836387, 7 pages, 2013. https://doi.org/10.1155/2013/836387
7. Fujiwara T, Yano Y, Hoshide S, Kanegae H, Kario K. Association of Cardiovascular Outcomes With Masked Hypertension Defined by Home Blood Pressure Monitoring in a Japanese General Practice Population. *JAMA Cardiol.* 2018;3(7):583–590. doi:10.1001/jamacardio.2018.1233
8. Sarah Cuschieri, Josanne Vassallo, Neville Calleja, Nikolai Pace, Julian Mamo, "The Effects of Socioeconomic Determinants on Hypertension in a Cardiometabolic At-Risk European Country", International Journal of Hypertension, vol. 2017, Article ID 7107385, 7 pages, 2017. https://doi.org/10.1155/2017/7107385
9. Xinyi Wu, Zhonghua Wang, "Role of Socioeconomic Status in Hypertension among Chinese Middle-Aged and Elderly Individuals", International Journal of Hypertension, vol. 2019, Article ID 6956023, 6 pages, 2019. https://doi.org/10.1155/2019/6956023

10. Veisani, Y., Jenabi, E., Nematollahi, S., Delpisheh, A., & Khazaei, S. (2019). The role of socio-economic inequality in the prevalence of hypertension in adults. *Journal of cardiovascular and thoracic research*, *11*(2), 116–120. https://doi.org/10.15171/jcvtr.2019.20
11. Philip, R., Beaney, T., Appelbaum, N. *et al.* Variation in hypertension clinical practice guidelines: a global comparison. *BMC Med* 19, 117 (2021). https://doi.org/10.1186/s12916-021-01963-0
12. Saklayen MG and Deshpande NV (2016) Timeline of History of Hypertension Treatment. *Front. Cardiovasc. Med.* 3:3. doi: 10.3389/fcvm.2016.00003
13. Devkota S, Dhungana RR, Pandey AR, Bista B, Panthi S, Thakur KK and Gajurel RM (2016) Barriers to Treatment and Control of Hypertension among Hypertensive Participants: A Community-Based Cross-sectional Mixed Method Study in Municipalities of Kathmandu, Nepal. *Front. Cardiovasc. Med.* 3:26. doi: 10.3389/fcvm.2016.00026
14. Sibomana, J.P., McNamara, R.L. & Walker, T.D. Patient, clinician and logistic barriers to blood pressure control among adult hypertensives in rural district hospitals in Rwanda: a cross-sectional study. *BMC Cardiovasc Disord* 19, 231 (2019). https://doi.org/10.1186/s12872-019-1203-3
15. Prenissl J, Manne-Goehler J, Jaacks LM, Prabhakaran D, Awasthi A, Bischops AC, et al. (2019) Hypertension screening, awareness, treatment, and control in India: A nationally representative cross-sectional study among individuals aged 15 to 49 years. PLoS Med 16(5): e1002801. https://doi.org/10.1371/journal.pmed.1002801
16. Schwartz, C.L., McManus, R.J. What is the evidence base for diagnosing hypertension and for subsequent blood pressure treatment targets in the prevention of cardiovascular disease?. *BMC Med* 13, 256 (2015). https://doi.org/10.1186/s12916-015-0502-5
17. Lee, H.Y., Lee, DJ., Seo, J. *et al.* Smartphone / smartwatch-based cuffless blood pressure measurement : a position paper from the Korean Society of Hypertension. *Clin Hypertens* 27, 4 (2021). https://doi.org/10.1186/s40885-020-00158-8
18. Emily P. Jolles, Raj S. Padwal, Alexander M. Clark, Branko Braam, "A Qualitative Study of Patient Perspectives about Hypertension", International Scholarly Research Notices, vol. 2013, Article ID 671691, 10 pages, 2013. https://doi.org/10.5402/2013/671691

19. Yoshida, T., Nishigaki, N., Saita, S. *et al.* Perspectives of patients and physicians regarding hypertensive management from an online survey for excellence: a subanalysis of the PARADOX study by physician categories. *Hypertens Res* 43, 431–441 (2020). https://doi.org/10.1038/s41440-019-0365-9

Chapter 2

Hypertension in Children and Adolescents: The Cases and the Solutions

Shivsanjeevi Sripathi

"...the salient feature of the absurd age I was at--an age which for all its alleged awkwardness, is prodigiously rich-- is that reason is not its guide, and the most insignificant attributes of other people always appear to be consubstantial with their personality. One lives among monsters and gods, a stranger to peace of mind. There is scarcely a single one of our acts from that time which we would not prefer to abolish later on. But all we should lament is the loss of the spontaneity that urged them upon us. In later life, we see things with a more practical eye, one we share with the rest of society; but adolescence was the only time when we ever learned anything."

–Proust

CONTENTS

2.1 The Normal Values 48
2.2 What Are The Figures? A Global Snapshot from the US to Africa 53
2.3 Grading and Diagnosis 60
2.4 The Quest for Links and Factors Involved 65
2.5 A neonatal Link 86
2.6 Hypertension and Sleep 89
2.7 The Effect of Seasons 96
2.8 Physical Activity 97
2.9 Final Points 104
2.10 References: 107

This chapter elucidates the normal values to blood pressure in children followed by diagnosing hypertension. This is followed by a snapshot of the global prevalence of this condition in children and adolescents. Several associated factors such as obesity, BMI, fructose consumption are also discussed. The importance of sleep and physical activity as outlined by research is presented followed by guidelines in managing hypertension in this age group.

2.1 THE NORMAL VALUES

According to the 4th Report on the Diagnosis, Evaluation and Treatment of High Blood Pressure in Children and Adolescents, Table 1 below summarizes blood pressure percentiles in boys as per height and age (Spagnolo et al, 2013: *Open Access: This article is published under license to BioMed Central Ltd. This is an Open Access article is distributed under the terms of the Creative Commons Attribution License (https://creativecommons.org/licenses/by/2.0)*, which permits unrestricted use, distribution, and reproduction in any medium, provided the original work is properly cited):

		Systolic (mmHg) percentile of height							Diastolic (mmHg) percentile of height						
Age (years)	**BP per-cen-tile**	**5th**	**10th**	**25th**	**50th**	**75th**	**90th**	**95th**	**5th**	**10th**	**25th**	**50th**	**75th**	**90th**	**95th**
1	90th	94	95	97	99	100	102	103	49	50	51	52	53	53	54
	95th	98	99	101	103	104	106	106	54	54	55	56	57	58	58
	99th	105	106	108	110	112	113	114	61	62	63	64	65	66	66
2	90th	97	99	100	102	104	105	106	54	55	56	57	58	58	59
	95th	101	102	104	106	108	109	110	59	59	60	61	62	63	63
	99th	109	110	111	113	115	117	117	66	67	68	69	70	71	71
3	90th	100	101	103	105	107	108	109	59	59	60	61	62	63	63
	95th	104	105	107	109	110	112	113	63	63	64	65	66	67	67
	99th	111	112	114	116	118	119	120	71	71	72	73	74	75	75
4	90th	102	103	105	107	109	110	111	62	63	64	65	66	66	67
	95th	106	107	109	111	112	114	115	66	67	68	69	70	71	71
	99th	113	114	116	118	120	121	122	74	75	76	77	78	78	79
5	90th	104	105	106	108	110	111	112	65	66	67	68	69	69	70
	95th	108	109	110	112	114	115	116	69	70	71	72	73	74	74
	99th	115	116	118	120	121	123	123	77	78	79	80	81	81	82
6	90th	105	106	108	110	111	113	113	68	68	69	70	71	72	72
	95th	109	110	112	114	115	117	117	72	72	73	74	75	76	76
	99th	116	117	119	121	123	124	125	80	80	81	82	83	84	84

7	90th	106	107	109	111	113	114	115	70	70	71	72	73	74	74
	95th	110	111	113	115	117	118	119	74	74	75	76	77	78	78
	99th	117	118	120	122	124	125	126	82	82	83	84	85	86	86
8	90th	107	109	110	112	114	115	116	71	72	72	73	74	75	76
	95th	111	112	114	116	118	119	120	75	76	77	78	79	79	80
	99th	119	120	122	123	125	127	127	83	84	85	86	87	87	88
9	90th	109	110	112	114	115	117	118	72	73	74	75	76	76	77
	95th	113	114	116	118	119	121	121	76	77	78	79	80	81	81
	99th	120	121	123	125	127	128	129	84	85	86	87	88	88	89
10	90th	111	112	114	115	117	119	119	73	73	74	75	76	77	78
	95th	115	116	117	119	121	122	123	77	78	79	80	81	81	82
	99th	122	123	125	127	128	130	130	85	86	86	88	88	89	90
11	90th	113	114	115	117	119	120	121	74	74	75	76	77	78	78
	95th	117	118	119	121	123	124	125	78	78	79	80	81	82	82
	99th	124	125	127	129	130	132	132	86	86	87	88	89	90	90
12	90th	115	116	118	120	121	123	123	74	75	75	76	77	78	79
	95th	119	120	122	123	125	127	127	78	79	80	81	82	82	83
	99th	126	127	129	131	133	134	135	86	87	88	89	90	90	91
13	90th	117	118	120	122	124	125	126	75	75	76	77	78	79	79
	95th	121	122	124	126	128	129	130	79	79	80	81	82	83	83
	99th	128	130	131	133	135	136	137	87	87	88	89	90	91	91
14	90th	120	121	123	125	126	128	128	75	76	77	78	79	79	80
	95th	124	125	127	128	130	132	132	80	80	81	82	83	84	84
	99th	131	132	134	136	138	139	140	87	88	89	90	91	92	92
15	90th	122	124	125	127	129	130	131	76	77	78	79	80	80	81
	95th	126	127	129	131	133	134	135	81	81	82	83	84	85	85
	99th	134	135	136	138	140	142	142	88	89	90	91	92	93	93
16	90th	125	126	128	130	131	133	134	78	78	79	80	81	82	82
	95th	129	130	132	134	135	137	137	82	83	83	84	85	86	87
	99th	136	137	139	141	143	144	145	90	90	91	92	93	94	94
17	90th	127	128	130	132	134	135	136	80	80	81	82	83	84	84
	95th	12	132	134	136	138	139	140	84	85	86	87	87	88	89
	99th	139	140	141	143	145	146	147	92	93	93	94	95	96	97

Table 2 below lists the blood pressure percentiles for girls by age and height (Spagnolo et al, 2013: *Open Access: This article is published under license to BioMed Central Ltd. This is an Open Access article is distributed under the terms of the Creative Commons Attribution License (https://creativecommons.org/licenses/by/2.0*), which permits unrestricted use, distribution, and reproduction in any medium, provided the original work is properly cited):

		Systolic (mmHg0 percentile of height							Diastolic (mmHg) percentile of height						
Age (years)	**BP per-cen-tile**	**5th**	**10th**	**25th**	**50th**	**75th**	**90th**	**95th**	**5th**	**10th**	**25th**	**50th**	**75th**	**90th**	**95th**
1	90th	97	97	98	100	101	102	103	52	53	53	54	55	55	56
	95th	100	101	102	104	105	106	107	56	57	57	58	59	59	60
	99th	108	108	109	111	112	113	114	64	64	65	65	66	67	67
2	90th	98	99	100	101	103	104	105	57	58	58	59	60	61	61
	95th	102	103	104	105	107	108	109	64	62	62	63	64	65	65
	99th	109	110	111	112	114	115	116	69	69	70	70	71	72	72
3	90th	100	100	102	103	104	106	106	61	62	62	63	64	64	65
	95th	104	104	105	107	108	109	110	65	66	66	67	68	68	69
	99th	111	111	113	114	115	116	117	73	73	74	74	75	76	76
4	90th	101	102	103	104	106	107	108	64	64	65	66	67	67	68
	95th	105	106	107	108	110	111	112	68	68	69	70	71	71	72
	99th	112	113	114	115	117	118	119	76	76	76	77	78	79	79
5	90th	103	103	105	106	107	109	109	66	67	67	68	69	69	70
	95th	107	107	108	110	111	112	113	70	71	71	72	73	73	74
	99th	114	114	116	117	118	120	120	78	78	79	79	80	81	81
6	90th	104	105	106	108	109	110	111	68	68	69	70	70	71	72
	95th	108	109	110	111	113	114	115	72	72	73	74	74	75	76
	99th	115	116	117	119	120	121	122	80	80	80	81	82	83	83
7	90th	106	107	108	109	111	112	113	69	70	70	71	72	72	73
	95th	110	111	112	113	115	116	116	73	74	74	75	76	76	77
	99th	117	118	119	120	122	123	124	81	81	82	82	83	84	84
8	90th	108	109	110	111	113	114	114	71	71	71	72	73	74	74
	95th	112	112	114	115	116	118	118	75	75	75	76	77	78	78
	99th	119	120	121	122	123	125	125	82	82	83	83	84	85	86
9	90th	110	110	112	113	114	116	116	72	72	72	73	74	75	75
	95th	114	114	115	117	118	119	120	76	76	76	77	78	79	79
	99th	121	121	123	124	125	127	127	83	83	84	84	85	86	87
10	90th	112	112	114	115	116	118	118	73	73	73	74	75	76	76
	95th	116	116	117	119	120	121	122	77	77	77	78	79	80	80
	99th	123	123	125	126	127	129	129	84	84	85	86	86	87	88
11	90th	114	114	116	117	118	119	120	74	74	74	75	76	77	77
	95th	118	118	119	121	122	123	124	78	78	78	79	80	81	81
	99th	125	125	126	128	129	130	131	85	85	86	87	87	88	89
12	90th	116	116	117	119	120	121	122	75	758	75	76	77	78	78
	95th	119	120	121	123	124	125	126	79	79	79	80	81	82	82

	99th	127	127	128	130	131	132	133	86	86	87	88	88	89	90
13	90th	117	118	119	121	122	123	124	76	76	76	77	78	79	79
	95th	121	122	123	124	126	127	128	80	80	80	81	82	83	83
	99th	128	129	130	132	133	134	135	87	87	88	89	89	90	91
14	90th	119	120	121	122	124	125	125	77	77	77	78	79	80	80
	95th	123	123	125	126	127	129	129	81	81	81	82	83	84	84
	99th	130	131	132	133	135	136	136	88	88	89	90	90	91	92
15	90th	120	121	122	123	125	126	127	78	78	78	79	80	81	81
	95th	124	125	126	127	129	130	131	82	82	82	83	84	85	85
	99th	131	132	133	134	136	137	138	89	89	90	91	91	92	93
16	90th	121	122	123	124	126	127	128	78	78	79	80	81	81	82
	95th	125	126	127	128	130	131	132	82	82	83	84	85	85	86
	99th	132	133	134	135	137	138	139	90	90	90	91	92	93	93
17	90th	122	122	123	125	126	127	128	78	79	79	80	81	81	82
	95th	125	126	127	129	130	131	132	82	83	83	84	85	85	86
	99th	133	133	134	136	137	138	139	90	90	91	91	92	93	93

In the case of adults, Ambulatory Blood Pressure Monitoring refers to monitoring the blood pressure over twenty-four hours to diagnose the condition. This can facilitate the identification of "white coat hypertension" (high office blood pressure values and normal Ambulatory Blood Pressure Monitoring values) and "masked hypertension" (normal office blood pressure values and elevated Ambulatory Blood Pressure Monitoring values). For children, the monitoring of 24-h ambulatory blood pressure by age and gender is represented in table 3 below (Spagnolo et al, 2013: *Open Access: This article is published under license to BioMed Central Ltd. This is an Open Access article is distributed under the terms of the Creative Commons Attribution License (https://creativecommons.org/licenses/by/2.0)*, which permits unrestricted use, distribution, and reproduction in any medium, provided the original work is properly cited):

Age (years)	Boys						Girls					
	Day			Night			Day			Night		
	75th	90th	95th	75th	90th	95th	75th	90th	95th	75th	90th	95th
5	116/76	120/79	123/81	99/59	103/62	106/65	114/77	118/80	121/82	100/81	105/66	108/69
6	116/76	121/79	124/81	100/59	105/63	108/66	155/77	120/80	122/82	101/61	106/65	110/68
7	117/76	122/80	125/82	101/60	106/64	110/67	116/77	121/80	123/82	102/60	107/65	111/67
8	117/76	122/80	125/82	102/60	108/64	111/67	117/76	122/80	124/82	103/60	108/64	112/67
9	118/76	123/80	126/82	103/60	109/64	112/67	118/76	122/80	125/82	103/59	109/64	112/6
10	119/76	124/80	127/82	104/60	110/64	113/67	119/76	123/79	126/81	104/59	110/64	113/67
11	121/76	126/80	129/82	105/60	111/64	115/67	120/76	120/79	127/81	105/59	110/63	114/66
12	123/76	128/80	132/82	107/60	113/64	116/67	121/76	125/80	128/82	105/59	110/63	114/66

13	126/76	131/80	1365/82	109/60	115/64	119/67	122/77	126/80	129/82	106/59	111/63	114/66
14	129/77	134/80	138/82	112/61	118/64	121/67	123/77	127/80	130/82	106/59	111/63	114/65
15	132/77	137/81	141/83	114/61	120/64	123/66	124/77	128/80	130/82	107/59	111/63	114/65
16	135/78	140/81	144/84	117/61	123/64	126/66	124/77	129/77	131/82	107/59	111/63	114/65

The 24-h ambulatory blood pressure by height and gender is represented below in table 4 below (Spagnolo et al, 2013: *Open Access: This article is published under license to BioMed Central Ltd. This is an Open Access article is distributed under the terms of the Creative Commons Attribution License (https://creativecommons.org/licenses/by/2.0*), which permits unrestricted use, distribution, and reproduction in any medium, provided the original work is properly cited):

Height (cm)	**Boys**						**Girls**					
	Day			**Night**			**Day**			**Night**		
	75th	**90th**	**95th**	**75th**	**90th**	**95th**	**75th**	**90th**	**95th**	**75th**	**90th**	**95th**
120	116/77	111/80	125/82	99/58	103/61	106/63	114/77	118/80	120/82	99/60	103/63	106/65
125	117/76	122/80	125/82	100/58	105/61	108/63	115/77	119/80	121/82	100/60	104/63	107/66
130	117/76	122/80	126/82	101/59	106/62	110/64	116/76	120/80	122/82	101/59	106/63	108/66
135	117/76	123/80	126/82	102/59	108/63	111/65	116/76	120/80	123/82	102/59	107/63	109/66
140	118/76	123/80	126/82	104/60	109/63	113/65	117/76	121/80	124/82	103/59	108/63	110/66
145	119/76	124/79	127/81	105/60	111/64	144/66	11876	123/80	125/82	103/59	109/63	112/66
150	120/76	125/79	128/81	106/60	112/64	116/66	119/76	124/80	127/82	104/59	110/63	113/66
155	122/76	127/79	130/81	107/60	113/64	117/66	121/76	125/80	128/82	106/59	111/63	114/66
160	124/76	129/79	133/81	108/60	144/64	188/66	122/76	126/80	129/82	106/59	111/63	114/66
165	126/76	132/80	135/82	110/60	116/64	119/66	123/77	127/80	130/82	107/59	112/63	114/66
170	128/77	134/80	138/82	112/61	117/64	121/66	124/77	128/80	131/82	108/61	112/67	115/71
175	130/77	136/81	140/83	113/61	119/64	112/66	125/78	129/81	131/82	109/59	113/63	115/66
180	132/77	138/81	142/83	115/61	120/64	124/66	N/A	N/A	N/A	N/A	N/A	N/A
185	134/78	140/81	144/84	116/61	122/64	125/66	N/A	N/A	N/A	N/A	N/A	N/A

When self-monitoring at home is followed, two measurements within a few minutes in the morning and in the evening for 3 consecutive days are warranted. The reference values of home blood pressure by height and gender are given below in table 5 (Spagnolo et al, 2013: *Open Access: This article is published under license to BioMed Central Ltd. This is an Open Access article is distributed under the terms of the Creative Commons Attribution License (https://creativecommons.org/licenses/by/2.0*), which permits unrestricted use, distribution, and reproduction in any medium, provided the original work is properly cited):

Height (cm)	Boys			Girls		
	N	50th	95th	*N*	50th	95th
120-129	23	105/64	119/76	36	101/64	119/74
130-139	51	108/64	101/77	51	103/64	120/76
140-149	39	110/65	125/77	61	105/65	122/77
150-159	41	112/65	126/78	71	108/66	123/77
160-169	45	115/65	123/78	148	110/66	124/78
170-179	91	117/66	132/78	46	112/66	125/79
180-189	57	121/67	134/79	7	114/67	128/80

2.2 WHAT ARE THE FIGURES? A GLOBAL SNAPSHOT FROM THE US TO AFRICA

On the basis of the US National Health and Nutrition Examination Survey (NHANES) data of 9117 children (8 to 12 years old) and 10156 adolescents (13 to 17 years old) between March 1999 and December 2018, Hardy and team (2021) assessed the trends of systolic BP (SBP) and diastolic BP (DBP). Analyses revealed that the age-adjusted mean SBP decreased from 102.4 mm Hg in 1999-2002 to 101.5 mm Hg in 2011-2014 followed by an increase to 102.5 mm Hg in 2015-2018in 8-12-year-olds. For the adolescents, the age-adjusted mean SBP decreased from 109.2 mm Hg in 1999-2002 to 108.4 mm Hg in 2011-2014 followed by no change in 2015-2018. For mean DBP, the former group displayed a lowering from 57.2 mm Hg in 1999-2002 to 51.9 mm Hg in 2011-2014 followed by an elevation to 53.2 mm Hg in 2015-2018. A lowering from 62.6mm Hg in 1999-2002 to 59.6 mm Hg in 2011-2014 and a spike to 60.8 mm Hg in 2015-2018 was observed.

Figure 1 below is illustrative of ***TOP:*** the trends in blood pressure in children (8 to 12 years old) and adolescents (13 to 17 years old) based on the 1999-2002 to 2015-2018 cycles of the National Health and Nutrition Examination Survey (Hardy et al, 2021: *This is an open access article distributed under the terms of the CC-BY license, which permits unrestricted use, distribution, and reproduction in any medium. You are not required to obtain permission to reuse this article content, provided that you credit the author and journal*):

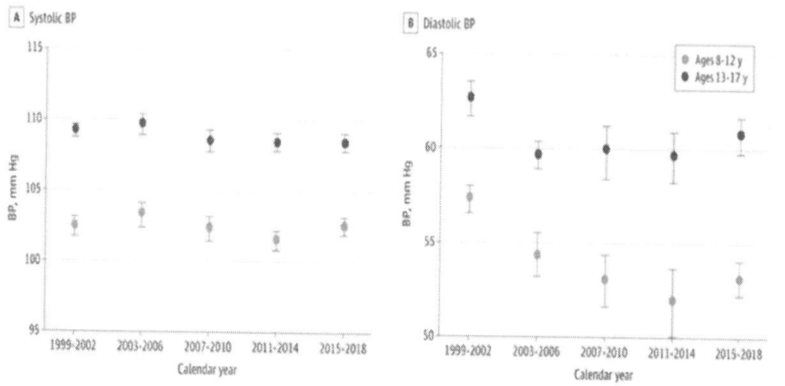

BOTTOM: Factors linked with high blood pressures in the children and adolescents (2015-2018) (Hardy et al, 2021: *This is an open access article distributed under the terms of the CC-BY license, which permits unrestricted use, distribution, and reproduction in any medium. You are not required to obtain permission to reuse this article content, provided that you credit the author and journal*):

Study participant characteristic	**NHANES cycle**[a]					***P* value**
	1999-2002	**2003-2006**	**2007-2010**	**2011-2014**	**2015-2018**	
Children aged 8-12 y						
NHANES sample size, No.	1981	1805	1808	1854	1669	NA
Age, mean (95% CI), y	10.5 (10.5-10.6)	10.5 (10.5-10.6)	10.5 (10.5-10.6)	10.5 (10.5-10.6)	10.5 (10.5-10.5)	.44
Sex						
Female	47.7 (45.1-50.3)	48.1 (44.7-51.9)	49.8 (47.0-52.6)	49.3 (46.0-52.7)	48.7 (45.2-52.2)	.53
Male	52.3 (49.7-54.9)	51.7 (48.1-55.3)	50.2 (47.4-53.0)	50.7 (47.3-54.0)	51.3 (47.8-54.8)	
Race/ethnicity						

Non-Hispanic White	59.1 (53.8-64.2)	57.6 (51.0-63.9)	56.7 (50.6-62.5)	54.7 (47.8-61.4)	49.7 (42.2-57.1)	.002
Non-Hispanic Black	16.3 (12.7-20.6)	15.9 (12.7-19.8)	13.8 (11.3-16.8)	13.5 (10.7-16.8)	13.7 (10.3-18.1)	
Hispanic	19.8 (15.4-25.1)	18.1 (14.4-22.6)	21.5 (16.8-27.0)	23.1 (17.9-29.3)	25.5 (19.9-32.0)	
Non-Hispanic Asian[b]	NA	NA	NA	4.3 (3.3-5.4)	4.7 (3.2-6.7)	
Non-Hispanic other	4.8 (3.6-6.3)	8.4 (6.0-11.7)	8.0 (6.2-10.4)	4.5 (3.4-5.9)	6.5 (4.9-8.5)	
BMI[c]						
Normal	65.6 (61.9-69.1)	62.5 (57.9-66.9)	60.5 (58.3-62.6)	61.9 (58.2-65.5)	61.6 (57.7-65.4)	.12
Overweight	16.8 (14.6-19.2)	18.1 (15.7-20.8)	18.1 (15.7-20.8)	17.6 (15.3-20.0)	17.5 (15.5-19.6)	
Obesity	17.6 (14.7-20.9)	19.4 (16.4-22.8)	21.5 (19.7-23.4)	20.5 (18.0-23.4)	20.9 (18.1-24.1)	
PIR[d]						
<1.30	35.2 (31.4-39.2)	29.5 (25.3-34.0)	33.0 (28.7-37.7)	36.3 (30.9-42.0)	30.0 (26.2-34.2)	.73
1.30-3.49	37.1 (33.0-41.4)	40.0 (35.8-44.3)	35.4 (31.8-39.2)	34.3 (30.3-38.5)	41.2 (37.3-45.3)	
≥3.50	27.7 (24.1-31.7)	30.5 (25.1-36.6)	31.6 (27.2-36.4)	29.5 (24.2-35.3)	28.7 (28.7-33.7)	
Adolescents aged 13-17 y						
NHANES sample size, No.	2889	2672	1522	1588	1485	NA
Age, mean (95% CI), y	15.5 (15.5-15.5)	15.5 (15.5-15.5)	15.5 (15.5-15.5)	15.5 (15.5-15.5)	15.5 (15.5-15.5)	.61
Sex						

Female	50.2 (47.4-53.0)	50.0 (47.5-52.6)	48.9 (45.8-52.1)	50.7 (47.2-54.3)	49.1 (46.1-52.2)	.74
Male	49.8 (47.0-52.6)	50.0 (47.4-52.5)	51.1 (47.9-54.2)	49.3 (45.7-52.8)	50.9 (47.8-53.9)	
Race/ethnicity						
Non-Hispanic White	60.1 (56.1-64.0)	64.5 (58.9-69.8)	59.8 (54.4-64.9)	54.0 (46.9-61.0)	53.3 (46.4-60.1)	.004
Non-Hispanic Black	14.1 (11.0-17.8)	14.5 (11.5-18.2)	14.9 (12.5-17.7)	15.0 (11.2-19.7)	13.9 (10.3-18.7)	
Hispanic	18.1 (14.1-23.1)	15.5 (12.4-19.2)	18.4 (14.2-23.4)	22.0 (17.4-27.5)	21.9 (16.6-28.2)	
Non-Hispanic Asian[b]	NA	NA	NA	5.3 (4.4-6.4)	4.6 (3.2-6.5)	
Non-Hispanic other	7.7 (5.7-10.3)	5.4 (3.8-7.7)	7.0 (5.1-9.4)	3.7 (2.6-5.3)	6.3 (4.7-8.5)	
BMI[c]						
Normal	68.5 (66.4-70.4)	65.3 (61.5-68.9)	65.0 (62.0-68.0)	63.1 (59.6-66.4)	59.2 (56.4-61.9)	<.001
Overweight	15.1 (13.5-16.8)	17.2 (14.9-19.9)	15.8 (13.8-18.2)	15.3 (13.2-17.6)	19.2 (17.5-21.0)	
Obesity	16.5 (15.2-17.8)	17.4 (14.9-20.3)	19.1 (16.3-22.3)	21.6 (18.6-25.0)	21.6 (18.8-24.6)	
PIR[d]						
<1.30	30.5 (27.6-33.6)	24.8 (21.8-28.1)	29.1 (25.6-32.7)	30.6 (25.3-36.4)	26.6 (22.3-31.4)	.42
1.30-3.49	36.2 (32.5-40.2)	38.1 (34.5-41.8)	35.3 (30.8-40.1)	38.6 (34.6-42.8)	43.0 (39.1-46.9)	
≥3.50	33.2 (29.6-37.1)	37.1 (32.6-41.9)	35.6 (30.3-41.3)	30.8 (25.6-36.7)	30.4 (25.5-35.8)	

Abbreviations: BMI, body mass index (calculated as weight in kilograms divided by height in meters squared); NA, not applicable; NHANES,

National Health and Nutrition Examination Survey; PIR, poverty-to-income ratio.

[a]Unless otherwise indicated, data are expressed as percentage (95% CI).

[b]Non-Hispanic Asian participants were not categorized in 1999-2002, 2003-2006, and 2007-2010 owing to small sample sizes.

[c]Normal indicates BMI of 5th percentile to less than 85th percentile; overweight, 85th percentile to less than 95th percentile; and obesity, 95th percentile or greater. Categories were based on age- and sex-specific growth charts developed by the Centers for Disease Control and Prevention in 2000. Owing to a small sample size, children with a BMI less than the 5th percentile are not presented.

[d]A PIR <1.30 indicates that the family income was below 130% of the poverty level.

Figure 2 below is illustrative of this outcome (© 2020 Daniel et al. *This is an open access article distributed under the terms of the Creative Commons Attribution License, which permits unrestricted use, distribution, and reproduction in any medium, provided the original author and source are credited*):

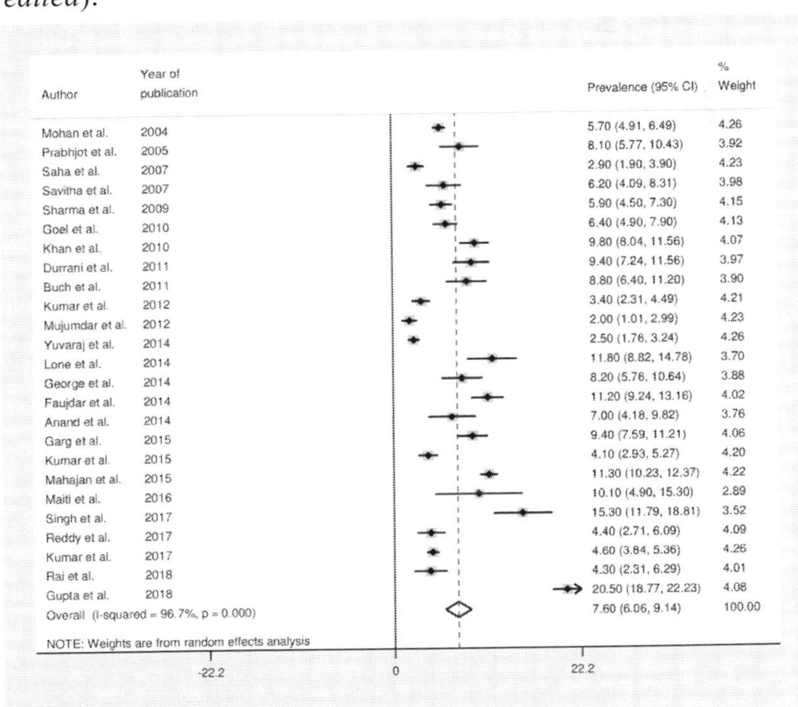

Given that an opportune detection of hypertension can circumvent several complications associated with this condition, researchers Daniel and team (2020) presented a systematic review across PubMed, Embase, IndMed, Cochrane library and Google Scholar (International Prospective Register of Systematic Reviews (PROSPERO) database number CRD42019132159) of the occurrence of hypertension in adolescents (10-19 years old). Subsequent analyses of 25 studies (inception to 1st March 2020) inclusive of 27,682 participants revealed that the pooled prevalence of the condition was 7.6%. According to a review in Clinical Hypertension, Ashraf et al (2020) summarized the overall trend with an elevation in BP by 19.2% in boys and 12.6% in girls in the US in the last decade. From India, 23% of 5–15-year-old children showed hypertension (systolic and/or diastolic) while 6–13 year-olds in Brazil (n=794) showed that 7% had hypertension. 4th-grade boys and girls in Japan recorded a prevalence of 15.9% and 15.8%, respectively. Looking at Africa, Ejike et al (2010) reported the trend in BP among adolescents in Kogi state, Nigeria in *BMC Pediatrics*. The sample set was inclusive of 13 year to 18-year-old adolescents of public secondary schools in Lokoja (urban) and Ajaokuta (semi-urban). The prevalence of point hypertension in the semi-urban area was 22.2% that increased with age: 15.1% at 13 years to 37.2% at 18 years. This trend was also observed in the urban dwellers from 5.3% at 13 years to 37.9% at 18 years (a seven-fold escalation).

Table 6 below is illustrative of the trend across adolescents in Nigerian localities (Ejike et al, 2010: *This article is published under license to BioMed Central Ltd. This is an Open Access article distributed under the terms of the Creative Commons Attribution License (http://creativecommons.org/licenses/by/2.0), which permits unrestricted use, distribution, and reproduction in any medium, provided the original work is properly cited*): *PHT: point-prehypertension and HT: point-hypertension*

AGE (Yrs)	SEMI-URBAN AREA									URBAN AREA								
	BOYS			GIRLS			TOGETHER			BOYS			GIRLS			TOGETHER		
	N	PHT (%)	HT (%)	N	PHT (%)	HT (%)	N	PHT (%)	HT (%)	N	PHT (%)	HT (%)	N	PHT (%)	HT (%)	N	PHT (%)	HT (%)
13	65	16.9	1.5	41	12.2	2.4	106	15.1	1.9	3	0.0	0.0	16	6.3	0.0	19	5.3	0.0
14	86	17.4	8.1	46	26.1	4.3	132	20.5	6.8	10	10.0	40.0	33	18.2	6.1	43	16.3	14.0
15	58	20.7	6.9	36	19.4	5.6	94	20.2	6.4	36	13.9	13.9	48	27.1	18.8	84	21.4	16.7
16	30	30.0	3.3	28	32.1	7.1	58	31.0	5.2	35	34.3	17.1	58	22.4	22.4	93	26.9	20.4
17	27	40.7	3.7	23	21.7	4.3	50	32.0	4.0	30	36.7	6.7	33	18.2	27.3	63	27.0	17.5
18	24	37.5	0.0	19	36.8	0.0	43	37.2	0.0	40	40.0	22.5	18	33.3	22.2	58	37.9	22.4
Total	290	23.1	4.8	193	20.7	4.1	483	22.2	4.6	154	29.2	16.9	206	21.8	18.0	360	25.0	17.5

No conspicuous differences were observed for the mean systolic blood pressures among both the genders who reported high blood pressures. For normotensive cases, except at 13 years of age, boys displayed a slightly higher trend in the mean systolic blood pressures. The mean diastolic blood pressures too displayed no evident variations between boys and girls with the highest values at 14 and 18 years for boys and 14 and 16 years for girls.

Figure 3 below depicts the trends in an adolescent population in Nigeria: ***TOP:*** Mean systolic blood pressures (Ejike et al, 2010: *This article is published under license to BioMed Central Ltd. This is an Open Access article distributed under the terms of the Creative Commons Attribution License (http://creativecommons.org/licenses/by/2.0), which permits unrestricted use, distribution, and reproduction in any medium, provided the original work is properly cited*):

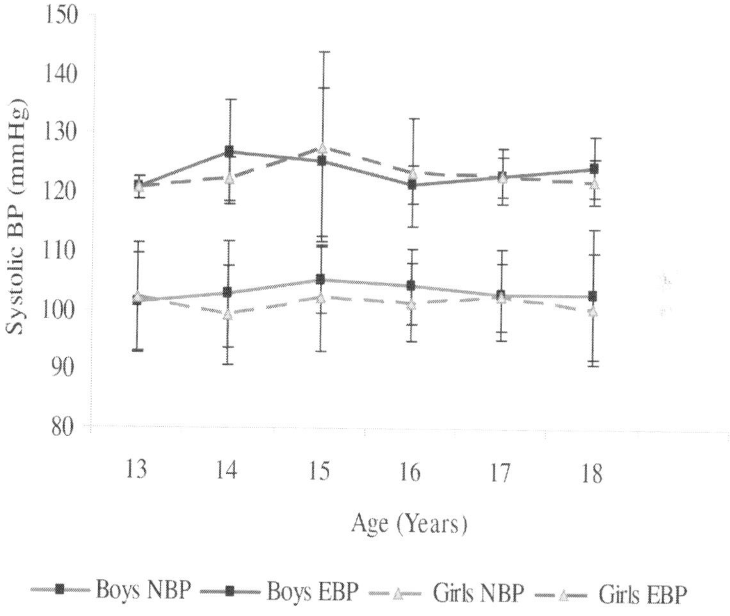

BOTTOM: Mean diastolic blood pressures (Ejike et al, 2010: *This article is published under license to BioMed Central Ltd. This is an Open Access article distributed under the terms of the Creative Commons Attribution License (http://creativecommons.org/licenses/by/2.0), which permits unrestricted use, distribution, and reproduction in any medium, provided the original work is properly cited*):

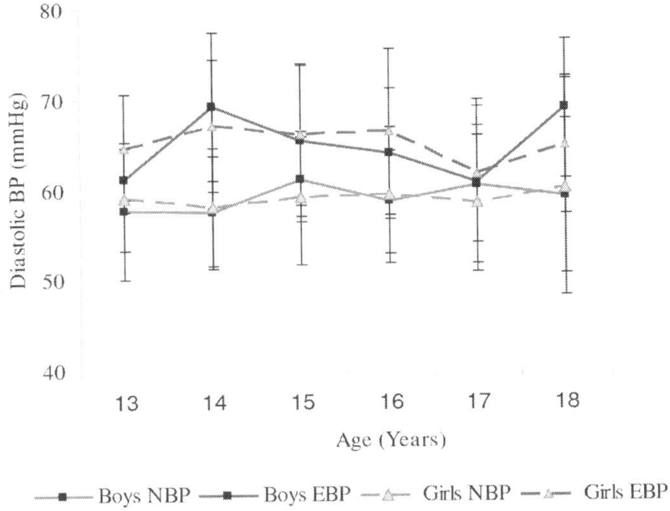

2.3 GRADING AND DIAGNOSIS

On the basis of the Fourth Report of National Blood Pressure Education Program Working Group on High Blood Pressure in Children and Adolescents (IV Report), Bonito and Di Sessa (2020) summarized the diagnosis of pediatric HTN:

Normal BP (BP < 90th percentile);

Pre-hypertension (BP ≥ 90th < 95th percentile);

Stage 1 hypertension (BP ≥ 95th < 99th +5 mmHg);

Stage 2 hypertension (BP ≥ 99th +5 mmHg)

Its subsequent update as per the new Clinical Practice Guidelines (CPG) of the American Academy of Pediatrics in 2017 is depicted below in table 7 (Bonito and Di Sessa, 2020: © 2020 by the authors. Licensee MDPI, Basel, Switzerland. This article is an open access article distributed under the terms and conditions of the Creative Commons Attribution (CC BY) license (*http://creativecommons.org/licenses/by/4.0/*):

	Systolic and/or Diastolic BP	**Systolic and/or Diastolic BP**
Age category	<13 years	≥13 years
Normal BP	<90th percentile	<120/80 mmHg
Elevated BP	≥90th <95th percentile or ≥120 <130/<80 mmHg	≥120 <130/<80 mmHg
Hypertension Stage 1	≥95th <95th percentile +12 or ≥130 <140/≥80 <90 mmHg	≥130 <140/≥80 <90 mmHg
Hypertension Stage 2	≥95th +12 or ≥140/90 mmHg	≥140/90 mmHg

Based on the National High Blood Pressure Education Program Working Group on High Blood Pressure in Children and Adolescents: The fourth report on the diagnosis, evaluation and treatment of high blood pressure in children and adolescents, Ashraf and team (2020) presented the same as in table 8 below. ***TOP:*** grading of pediatric hypertension (HTN) (Ashraf et al, 2020: Open Access: This article is licensed under a Creative Commons Attribution 4.0 International License, which permits use, sharing, adaptation, distribution and reproduction in any medium or format, as long as you give appropriate credit to the original author(s) and the source, provide a link to the Creative Commons *licence, and indicate if changes were made. The images or other third party material in this article are included in the article's Creative Commonslicence, unless indicated otherwise in a credit line to the material. If material is not included in the article's Creative Commons licence and your intended use is not permitted by statutory regulation or exceeds the permitted use, you will need to obtain permission directly from the copyright holder. To view a copy of this licence, visit http://creativecommons.org/licenses/by/4.0/. The Creative Commons Public Domain Dedication waiver (http://creativecommons.org/publicdomain/zero/1.0/) applies to the data* *made available in this article, unless otherwise stated in a credit line to the data*):

BP/HTN	Age 1–<13 Years	Age>13 Years
Normal BP	<90th percentile	<120/<80 mmHg
Elevated BP	≥90th percentile to <95th percentile or 120/80 mmHg to <95th percentile (whichever is lower)	120/<80–129/<80 mmHg
Stage 1HTN	≥95th percentile to <95th percentile+ 12 mmHg or 130/80–139/89 mmHg (whichever is lower)	130/80–139/89 mmHg
Stage 2 HTN	≥95th percentile + 12 mmHg or≥ 140/90 mmHg (whichever is lower)	≥140/90 mmHg

BOTTOM: The causes of hypertension in children (Ashraf et al, 2020: Open Access: This article is licensed under a Creative Commons Attribution 4.0 International License, which permits use, sharing, adaptation, distribution and reproduction in any medium or format, as long as you give appropriate credit to the original author(s) and the source, provide a link to the Creative Commons *licence, and indicate if changes were made. The images or other third party material in this article are included in the article's Creative Commonslicence, unless indicated otherwise in a credit line to the material. If material is not included in the article's Creative Commons licence and your intended use is not permitted by statutory regulation or exceeds the permitted use, you will need to obtain permission directly from the copyright holder. To view a copy of this licence, visit http://creativecommons.org/licenses/by/4.0/. The Creative Commons Public Domain Dedication waiver (http://creativecommons.org/publicdomain/zero/1.0/) applies to the data made available in this article, unless otherwise stated in a credit line to the data*):

Age	Common Causes
Newborn	Renal artery thrombosis or embolus, Renal vein thrombosis, Congenital renal malformations, Coarctation of aorta, Renal artery stenosis, Bronchopulmonary dysplasia
Infancy to 6 years	Renal parenchymal disease, Renal artery stenosis, Coarctation of the aorta, Medications (corticosteroids, albuterol, pseudoephedrine), Endocrine causes
6–10 years	Renal parenchymal disease, Renal artery stenosis, Primary hypertension, Endocrine causes.

Adoles-cence	Primary hypertension, White coat hypertension, Renal parenchymal disease, Substance abuse (cocaine, amphetamines, methamphetamines, phencyclidine, methylphenidate, caffeine), Teen pregnancy, Endocrine causes.

A slew of factors considered history require dedicated analyses such as birth weight, family history, events during gestation, food intake, sceen watching time, physical activity and diet. Further, general appearance factors inclusive of the general growth, BMI, pulse rate along with unusual manifestations such as excessive hair, rash, swellings, weakness, etc. also require scrutiny. In the event of congenital heart disease or malignancy or recurrent urinary tract infections or organ transplant or prematurity, screening of blood pressure is warranted.

Table 9 below capitulates the values the crossing of which warrants investigation (Ashraf et al, 2020: :

Age (Years)	**Boys**		**Girls**	
	SBP (mmHg)	**DBP (mmHg)**	**SBP (mmHg)**	**DBP (mmHg**
1	98	52	98	54
2	100	55	101	58
3	101	58	102	60
4	102	60	103	62
5	103	63	104	64
6	105	66	105	67

7	106	68	106	68
8	107	69	107	69
9	107	70	108	71
10	108	72	109	72
11	110	74	111	74
12	113	75	114	75
≥13	120	80	120	80

Table 10 below summarizes the aspects of medical history to be recorded in children and adolescents with hypertension (Spagnolo et al, 2013: *Open Access: This article is published under license to BioMed Central Ltd. This is an Open Access article is distributed under the terms of the Creative Commons Attribution License (https://creativecommons.org/licenses/by/2.0*), which permits unrestricted use, distribution, and reproduction in any medium, provided the original work is properly cited):

Family history	
•	Hypertension
•	Cardiovascular and cerebrovascular disease
•	Diabetes mellitus
•	Dyslipidemia
•	Obesity
•	Hereditary renal disease (Polycystic kidney disease)
•	Hereditary endocrine disease (pheochromocytoma, adrenal hyperplasia, multiple endocrine neoplasia , von Hippel-Lindau)
•	Syndromes associated with hypertension (neurofibromatosis)
Clinical history	
•	**Perinatal history**: birth weight, gestational age, oligohydramnios, anoxia, umbilical artery catheterization
•	**Previous history**: urinary tract infection, renal or urological disease, cardiac, endocrine (including diabetes) or neurological disease, growth retardation

•	**Symptoms suggestive of secondary hypertension**: dysuria, thirst/polyuria, nocturia, hematuria, edema, weight loss, failure to thrive, palpitations, sweating, fever, pallor, flushing, cold extremities, intermittent claudication, virilization, primary amenorrhea, male pseudohermaphroditism
•	**Symptoms suggestive of target organ damage**: headache, epistaxis, vertigo, visual impairment, facial palsy, fits, strokes, dyspnea
•	**Sleep history**: snoring, apnea, daytime somnolence
•	**Risk factor history**: low physical exercise level, incorrect dietary habits, smoking, alcohol, licorice
•	**Drug intake**: cyclosporine, tacrolimus, tricyclic anti-depressants, antipsychotics, decongestants, oral contraceptives, illegal drugs
•	**Pregnancy**

2.4 THE QUEST FOR LINKS AND FACTORS INVOLVED

Along with the pattern of hypertension in "a community based prospective, observational study", the factors associated with this medical condition were assessed in 6-16-year-old apparently healthy school children (total: 3,302=1,658 boys and 1,644 girls) in Secunderabad, India. The classification of hypertension was on the basis of the fourth report on the diagnosis, evaluation, and treatment of high blood pressure in children and adolescents i.e. "average systolic blood pressure and or diastolic blood pressure that is >95 percentile for gender, age, and height on ≥3 occasions".

Both systolic blood pressure (SBP) and diastolic blood pressure (DBP) displayed a positive association with age. Across all the ages the BP in male students was higher by 1–2 mm of Hg vs. that of their female counterparts. The plausible explanation is the increased heaviness and height in boys vs. girls at that age. For 6–10 year-olds, the SBP rose by 1–2 mm of Hg/year while this increase was 2–3 mm of Hg/year for adolescence (>11 years). The DBP spiked by 4–5 mm of Hg in 6–10 year-olds while the spike was 8–10 mm of Hg in adolescence. The physical and hormonal changes are plausible factors behind these spikes.

Height was found to be related to BP and serves as an independent variable for BP: 3–5 mm of Hg increase for every 10 cm increase in height irrespective of the weight or height. Further, the mean SBP and DBP increased by 3–4 mm of Hg up to 130 cm-tall boys and girls with the increase reaching 4–5 mm of Hg when the height exceeded 130 cm. In terms of weight, the DBP increased by 2.5–3 mm of Hg for every weight group up to 60 kg. Both SBP and DBP were 2–3 mm of Hg higher in students with a family history of hypertension (both genders). Overall, the researchers concluded "***body weight and height***" are the predominant determinants of BP in this age group.

(Sayeemuddin et al, 2015)

Table 11 below is illustrative of the factors influencing BP in 6-16-year-olds: ***TOP***: Correlation with height (*Copyright: © 2015 Sayeemuddin, Sharma, Pandita, Sultana and Shastri. This is an open-access article distributed under the terms of the* ***Creative Commons Attribution License (CC BY)****. The use, distribution or reproduction in other forums is permitted, provided the original author(s) or licensor are credited and that the original publication in this journal is cited, in accordance with accepted academic practice. No use, distribution or reproduction is permitted which does not comply with these terms*):

Height (cm)	Males			Females		
	No. of cases	Mean	SD	No. of cases	Mean	SD
MEAN SYSTOLIC BLOOD PRESSURE						
100–110	4	96.33	1.41	6	95.33	1.76
110–120	194	99.57	3.43	242	97.10	2.60
120–130	350	100.54	2.61	344	99.71	2.11
130–140	258	102.34	2.00	224	101.74	1.97
140–150	250	105.46	1.63	278	105.67	1.83
150–160	276	108.63	1.69	504	109.69	2.20
160–170	308	113.61	2.13	46	112.87	2.18
170–180	18	116.37	1.16	–	–	–
MEAN DIASTOLIC BLOOD PRESSURE						
100–110	4	56.67	1.89	6	57.78	1.69
110–120	194	60.39	3.28	242	58.41	2.19
120–130	350	62.64	1.45	344	61.21	1.47
130–140	258	64.64	1.08	224	62.84	1.34
140–150	250	65.91	0.98	278	63.88	1.62

150–160	276	68.47	1.20	504	67.47	2.19
160–170	308	72.56	1.76	46	70.29	1.69
170–180	18	73.85	1.19	–	–	–

BOTTOM: Correlation with weight (*Copyright:*):

Weight (kg)	**Males**			**Females**		
	No. of cases	**Mean**	**SD**	**No. of cases**	**Mean**	**SD**
MEAN SYSTOLIC BLOOD PRESSURE						
15–20	112	98.33	2.87	178	96.85	2.59
20–25	428	100.53	2.80	424	99.49	2.46
25–30	226	102.48	2.07	174	101.73	2.33
30–35	204	104.69	1.97	194	104.21	2.06
35–40	154	106.76	1.51	156	106.92	1.95
40–45	188	108.76	1.73	254	108.31	1.50
45–50	146	112.04	1.97	248	111.56	1.94
50–55	188	114.62	1.97	16	113.17	1.11
55–60	12	115.00	1.10	–	–	–
MEAN DIASTOLIC BLOOD PRESSURE						
15–20	112	59.31	2.95	178	58.03	1.79
20–25	428	62.34	1.95	424	60.88	1.84
25–30	226	64.25	0.91	174	62.74	1.16
30–35	204	65.37	0.81	194	63.24	1.50
35–40	154	67.00	0.75	156	65.06	1.29
40–45	188	68.79	0.67	254	66.01	1.47
45–50	146	70.95	1.13	248	69.38	1.31
50–55	188	73.60	1.04	16	72.25	0.79
55–60	12	75.33	0.60	–	–	–

Analyses of 722 children (7-10 years old)of 13 public elementary schools in Vitória, state of Espírito Santo, Brazilwas done by CamilaBrandão de Souzaand team (2017). The basis of the longitudinal study was: normal (BP < 90th percentile), borderline (BP between the 90th and 95th percentiles or if the BP was more than 120/80 mmHg, always < 90th percentile up to < 95th percentile) and high (95th percentile).

The researchers reported the highest systolic values at ages 9 and 10 at 103.9 mmHg ±11.2 mmHg and 104.3 mmHg ± 11.7 mmHg, respectively. In terms of nutritional status, overweight and obese children showed the highest systolic BP at 105.1 mmHg ± 10.6 mmHg and 106.3 mmHg ± 10.8 mmHg, respectively. This trend was on the same lines for the diastolic BP at ages 9 and 10 at 63.2 mmHg ± 10.1 mmHg and 62.4 mmHg ± 8.1 mmHg, respectively and in overweight and obese children at 64.2 mmHg ± 8, 2 mmHg and 65.0 mmHg ± 9.5 mmHg, respectively.

Table 12 below displays the trend in BP across the selected children in Brazil (CamilaBrandão de Souza et al, 2017:):

Key: *Analysis of variance; †Student's t test for independent samples, SBP: systolic blood pressure; DBP: diastolic blood pressure; SD: standard deviation

Characteristics	**SBP (mmHg)**		**DBP (mmHg)**	
	Mean ± SD	**p-value**	**Mean ± SD**	**p-value**
Age				
7	100.3 ± 10.8[a]	< 0.001*	61.5 ± 10.5[ab]	0.040*
8	101.1 ± 10.9[a]		60.8 ± 9.5[a]	
9	103.9 ± 11.2[b]		63.2 ± 10.1[b]	
10	104.3 ± 11.7[b]		62.4 ± 8.1[b]	
Gender				
Male	102.6 ± 11.7	0.237†	61.5 ± 10.3	0.159†
Female	101.9 ± 10.9		62.2 ± 9.3	
Ethnicity/race				

White	102.7 ± 10.6[a]	0.047*	63.0 ± 10.3	0.146*
Black	102.5 ± 11.0[a]		61.5 ± 9.0	
Brown	101.6 ± 11.5[a]		61.6 ± 9.6	
Asian/indigenous	107.6 ± 12.9[b]		63.0 ± 10.5	
Nutritional status				
Normal weight	100.3 ± 11.2[a]	< 0.001*	60.5 ± 9.9[a]	< 0.001*
Low weight	99.5 ± 9.3[a]		59.3 ± 8.3[a]	
Overweight	105.1 ± 10.6[b]		64.2 ± 8.2[b]	
Obesity	106.3 ± 10.8[b]		65.0 ± 9.5[b]	
Height deficit				
No	97.5 ± 16.6	0.189†	60.0 ± 14.4	0.210†
Yes	102.2 ± 11.2		62.0 ± 9.7	
Total	102.2 ± 11.3		61.9 ± 9.7	

8.1% of the children showed elevated BP at the first instance that was monitored for two more instances with 3.2% in the second reassessment and 2.1% retained the high BP the third time. The occurrence of BP was higher by a factor of five in obese children vs. that of the ones with regular weights showing the relevance of nutritional status in hypertension.

A large-scale analysis of 40,105 boys and 38,009 girls in the "Shanghai children and adolescents disease risk factor study" (SCADRFS) scrutinized the extent of hypertension across 60 schools in 9 urban districts and 50 schools in 10 rural districts of Shanghai, China. 22.6% of the children were found to be overweight or obese stage (on the basis of the BMI) that displayed a positively strong association with the risk of hypertension.

Figure 4 below is illustrative of the higher occurrence of high BP in obese children vs. those with regular weights (CamilaBrandão de Souza et al, 2017: *This is an Open Access article distributed under the terms of the Creative Commons Attribution License, which permits unrestricted use, distribution, and reproduction in any medium, provided the original work is properly cited*):

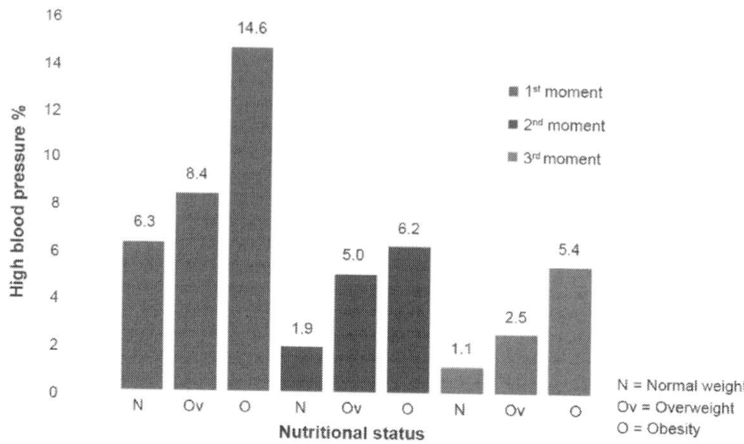

The systolic and diastolic blood pressures were elevated in the overweight (OW) and obese (OB) groups with waist circumference (WC) emerging as a more predominant risk factor over BMI. Table 13 below depicts these findings; TOP: Elevated blood pressure in OW and OB children: *BMI, body mass index; WC, waist circumference; DBP, diastolic blood pressure; SBP, systolic blood pressure*(Lu et al, 2013: *This article is published under license to BioMed Central Ltd. This is an Open Access article distributed under the terms of the Creative Commons Attribution License (http://creativecommons.org/licenses/by/2.0), which permits unrestricted use, distribution, and reproduction in any medium, provided the original work is properly cited*):

	BMI				WC			
	Normal weight	**Overweight**	**Obese**	***P*value**	**Normal weight**	**Overweight**	**Obese**	***P*-value**
	(*n*=60465)	**(*n*=10252)**	**(*n*=7397)**		**(*n*=62946)**	**(*n*=9109)**	**(*n*=5884)**	
Age (years)	11.3±3.1 (7,20)	11.2±3.0 (7,20)	10.5±2.8 (7,20)	<0.001	11.2=3.1 (7,20)	11.1±2.9 (7,20)	11.5±3.0 (7,20)	<0.001
Sex								
Boys	28652 (71.4%)	6568 (16.4%)	4885 (12.2%)	<0.001	31414 (78.5%)	5210 (13.0%)	3407 (8.5%)	<0.001
Girls	31813 (83.7%)	3684 (9.7%)	2512 (6.6%)		31532 (83.2%)	3899 (10.3%)	2477 (6.5%)	
Residence								
Urban	35067 (74.6%)	6865 (14.6%)	5091 (10.8%)	<0.001	36525 (77.9%)	6184 (13.2%)	4168 (8.9%)	<0.001
Rural	25398 (81.7%)	3387 (10.9%)	2306 (7.4%)		26421 (85.1%)	2925 (9.4%)	1716 (5.5%)	
Height (cm)	147±16 (106,196)	150±15 (110,198)	149±15 (84,196)	<0.001	147±16 (84,196)	150±14 (85,191)	155±15 (117,198)	<0.001
Weight (kg)	38.8±12.5 (13.9,86.7)	50.3±15.3 (18.6,105.5)	57.6±19.1 (17.9,139.1)	<0.001	38.9±12.6 (13.9,101.6)	50.7±13.8 (21.0,105.4)	63.2±18.5 (20.0,139.1)	<0.001
BMI(kg/m^2)	17.3±2.3 (8.6,30.7)	21.8±2.5 (11.9,27.8)	25.2±3.9 (12.1,63.5)	<0.001	17.5±2.6 (8.6,48.9)	21.9±2.7 (12.1,63.5)	25.7±3.8 (8.7,47.3)	<0.001
WC (cm)	60±7 (26,110)	70±9 (23,101)	78±12 (35,131)	<0.001	60±7 (23,80)	72±6 (58,87)	84±9 (64,131)	<0.001
SBP (mmHg)	102±12 (62,180)	107±13 (70,180)	110±15 (65,180)	<0.001	102±12 (62,180)	107±13 (65,180)	112±14 (70,180)	<0.001
DBP (mmHg)	65±9 (40,137)	67±9 (40,120)	69±10 (40,110)	<0.001	65±9 (40,137)	67±9 (40,120)	70±10 (40,110)	<0.001

Across-sectional study of 13597 six-twelve children in urban and rural counties in Chongqing in Southwest of China (XiaohuaLiang et al, 2020) scrutinized the extent and factors linked to childhood hypertension. Interestingly, the children dwelling in rural areas displayed higher heart rate, HDL and LDL among other factors while the weight and BMI were lower. Table 14 below summarizes the findings (Copyright © 2020 Xiaohua Liang et al. *This is an open access article distributed under the Creative Commons Attribution License, which permits unrestricted use, distribution, and reproduction in any medium, provided the original work is properly cited*):

Key: BMI, body mass index; FBG, fasting blood glucose; TC, total cholesterol; TG, triglyceride; HDL-C, high-density lipoprotein cholesterol; LDL-C, low-density lipoprotein cholesterol; WBC, white blood cell; RBC, red blood cell; HGB, hemoglobin; HCT, hematocrit; MCV, mean corpuscular volume; MCH, mean corpuscular hemoglobin; MCHC, mean corpuscular hemoglobin concentration; PLT, platelet; WSCR, WBC small cell ratio; WMCR, WBC middle cell ratio or mixed cell percent; WSCC, WBC small cell count or lymphocyte count; WMCC, WBC middle cell count or mixed cell count; WLCC, WBC large cell count or neutrophil count; RDWSD, red blood cell distribution width; PDW, platelet distribution width; MPV, mean platelet volume; P-LCR, platelet-large cell ratio

Variables	**Total**	**Urban vs rural**		
		Urban (n=5995)	***Rural (n=7602)***	***P-value***
Anthropometric measures				
Gender, male (n (%))	7060 (51.92%)	3098 (51.68%)	3962 (52.12%)	0.609
Age, y	9.24±1.76	9.05±1.76	9.39±1.74	<0.001
BMI, kg/m2	17.35±3.05	17.53±3.02	17.22±3.07	<0.001
Height, cm	134.51±11.5	134.66±11.44	134.39±11.56	0.182
Weight, kg	32.01±9.49	32.40±9.53	31.7±9.45	<0.001
Waist circumference, cm	58.12±8.49	59.17±8.73	57.3±8.21	<0.001
Heart rate, n/min	95.28±12.72	94.95±12.71	95.54±12.72	0.007
Serum biochemical index				
FBG, mmol/L	4.14±0.62	3.94±0.50	4.90±0.46	<0.001

TC, mmol/L	3.49±0.66	3.34±0.57	4.10±0.63	<0.001
TG, mmol/L	0.97±0.58	0.96±0.58	1.03±0.57	<0.001
HDL-C, mmol/L	1.22±0.27	1.19±0.26	1.35±0.28	<0.001
LDL-C, mmol/L	1.72±0.54	1.56±0.39	2.35±0.60	<0.001
Blood cell composition measures				
WBC, $n \times$ 10^9/L	7.12±1.79	7.12±1.77	7.16±1.85	0.480
RBC, $n \times$ 10^12/L	4.59±0.39	4.52±0.35	4.90±0.40	<0.001
HGB, g/L	126.42±9.98	124.31±8.56	134.73±10.82	<0.001
HCT, %	37.66±3.11	36.74±2.43	41.29±2.78	<0.001
MCV, fL	81.91±6.44	81.37±5.98	84.01±7.66	<0.001
MCH, pg	27.65±1.95	27.64±1.94	27.69±1.96	0.390
MCHC, g/L	336.58±14.98	338.99±15.08	327.11±10.06	<0.001
PLT, $n \times$ 10^9/L	287.48±67.49	288.53±67.42	283.35±67.63	0.020
WSCR, %	36.11±8.25	35.59±8.03	38.15±8.80	<0.001
WMCR, %	7.05±2.75	7.41±2.87	5.64±1.60	<0.001
WSCC, $n \times$ 10^9/L	2.52±0.66	2.47±0.63	2.70±0.73	<0.001
WMCC, $n \times$ 10^9/L	0.51±0.34	0.53±0.28	0.44±0.48	<0.001
WLCC, $n \times$ 10^9/L	4.08±1.49	4.12±1.48	3.90±1.50	<0.001
RDWSD, %	34.59±4.82	33.40±4.23	39.27±4.09	<0.001
PDW, fL	13.47±2.37	13.48±2.32	13.39±2.57	0.207
MPV, fL	10.68±1.30	10.57±1.32	11.11±1.15	<0.001
P-LCR, %	32.67±8.65	32.66±8.86	32.69±7.76	0.910
Socioeconomic index				
Father's education level, y				
~9	6856 (51.75%)	2190 (37.80%)	4666 (62.58%)	<0.001
~12	4915 (37.10%)	2787 (48.11%)	2128 (28.54%)	
~15	1390 (10.49%)	755 (13.03%)	635 (8.52%)	
>15	88 (0.66%)	61 (1.05%)	27 (0.36%)	
Father's occupation				
Manager	933 (7.02%)	389 (6.70%)	544 (7.26%)	<0.001
Worker	4016 (30.21%)	1733 (29.84%)	2283 (30.49%)	
Technicist/researcher	571 (4.29%)	380 (6.54%)	191 (2.55%)	
Farmer	4593 (34.55%)	1686 (29.03%)	2907 (38.82%)	
Others	3182 (23.93%)	1619 (27.88%)	1563 (20.87%)	
Mother's occupation				
Manager	605 (4.55%)	241 (4.12%)	364 (4.88%)	<0.001
Worker	3538 (26.58%)	1666 (28.45%)	1872 (25.11%)	

Technicist/researcher	239 (1.80%)	132 (2.25%)	107 (1.44%)	
Farmer	4179 (31.40%)	2111 (36.05%)	2068 (27.74%)	
Others	4750 (35.68%)	1705 (29.12%)	3045 (40.84%)	
Income, RMB				
~500	738 (6.25%)	92 (2.05%)	646 (8.83%)	<0.001
~1000	1395 (11.82%)	235 (5.23%)	1160 (15.86%)	
~2000	2327 (19.72%)	627 (13.96%)	1700 (23.25%)	
~3000	2750 (23.30%)	1051 (23.41%)	1699 (23.24%)	
>3000	4592 (38.91%)	2485 (55.35%)	2107 (28.82%)	
Living with grandparents				
No	10060 (84.76%)	3992 (88.42%)	6068 (82.51%)	<0.001
Yes	1809 (15.24%)	523 (11.58%)	1286 (17.49%)	
People live with child				
1	583 (4.98%)	154 (3.43%)	429 (5.94%)	<0.001
2~3	6114 (52.22%)	2595 (57.81%)	3519 (48.74%)	
4	5012 (42.80%)	1740 (38.76%)	3272 (45.32%)	
Medical insurance				
No	1790 (15.14%)	383 (8.50%)	1407 (19.24%)	<0.001
Yes	10030 (84.86%)	4124 (91.50%)	5906 (80.76%)	
Perinatal measures				
Gestational hypertension				
No	11299 (98.47%)	4366 (98.76%)	6933 (98.30%)	0.052
Yes	175 (1.53%)	55 (1.24%)	120 (1.70%)	
Birth weight, g				
~3000	3181 (26.86%)	1133 (25.16%)	2048 (27.91%)	<0.001
~3600	5051 (42.65%)	2058 (45.70%)	2993 (40.78%)	
>3600	3610 (30.48%)	1312 (29.14%)	2298 (31.31%)	
Breastfeeding, month				
0~3	2845 (24.40%)	1420 (31.77%)	1425 (19.81%)	<0.001
4~10	5555 (47.64%)	2223 (49.74%)	3332 (46.33%)	
>10	3261 (27.97%)	826 (18.48%)	2435 (33.86%)	
Father with obesity				
No	9839 (83.89%)	3703 (82.79%)	6136 (84.56%)	0.011
Yes	1890 (16.11%)	770 (17.21%)	1120 (15.44%)	
Mother with obesity				
No	10505 (89.57%)	4017 (89.73%)	6488 (89.48%)	0.670
Yes	1223 (10.43%)	460 (10.27%)	763 (10.52%)	

Physical activity				
Weekday, min/day				
~10	6321 (53.41%)	2246 (49.87%)	4075 (55.59%)	<0.001
>10	5514 (46.59%)	2258 (50.13%)	3256 (44.41%)	
Weekend, min/day				
~55	7782 (65.77%)	2795 (62.11%)	4987 (68.01%)	<0.001
>55	4051 (34.23%)	1705 (37.89%)	2346 (31.99%)	
Total, min/week				
~150	6210 (52.59%)	2170 (48.32%)	4040 (55.21%)	<0.001
>150	5599 (47.41%)	2321 (51.68%)	3278 (44.79%)	

The occurrence of hypertension was 13.75% in the children with the following factors influencing its occurrence as illustrated in table 15 below(Copyright © 2020 Xiaohua Liang et al. *This is an open access article distributed under the Creative Commons Attribution License, which permits unrestricted use, distribution, and reproduction in any medium, provided the original work is properly cited*):

Key: [a]A total of 348 subjects having missing values in father's education level; [b]A total of 302 subjects having missing values in father's occupation. [c]A total of 286 subjects having missing values in mother's occupation. [d]A total of 1795 subjects having missing values in parents' income. [e]A total of 1728 subjects having missing values in living with grandparents. [f]A total of 1888 subjects having missing values in people live with child. [g]A total of 1777 subjects having missing values in medical insurance. [h]A total of 2123 subjects having missing values in gestational hypertension. [i]A total of 1755 subjects having missing values in birth weight. [j]A total of 1936 subjects having missing values in breastfeeding. [k]A total of 1868 subjects having missing values in father with obesity. [l]A total of 1869 subjects having missing values in mother with obesity. [m]A total of 1762 subjects having missing values in weekday activity. [n]A total of 1764 subjects having missing values in weekend activity. [o]A total of 1788 subjects having missing values in total activity

Variables	Total		Urban		Rural		Difference (rural vs urban)	P-value
	Prevalence		*Prevalence*		*Prevalence*			
Total	13.75% (13.17%, 14.34%)		9.02% (8.31%, 9.78%)		17.47% (16.62%, 18.34%)		8.44% (7.32%, 9.56%)	<0.001
Anthropometric measures								
Gender								
Male	13.50% (12.71%, 14.32%)	0.385	8.72% (7.75%, 9.76%)	0.388	17.24% (16.07%, 18.45%)	0.581	8.52% (6.98%, 10.06%)	<0.001
Female	14.01% (13.18%, 14.88%)		9.35% (8.32%, 10.47%)		17.72% (16.49%, 19.00%)		8.37% (6.73%, 10.00%)	<0.001
Age, y								
6	16.73% (14.92%, 18.66%)	<0.001	14.15% (11.83%, 16.72%)	<0.001	19.50% (16.75%, 22.49%)	<0.001	5.36% (1.67%, 9.04%)	0.004
7	15.66% (14.26%, 17.15%)		10.91% (9.23%, 12.79%)		20.35% (18.15%, 22.69%)		9.44% (6.61%, 12.26%)	<0.001
8	13.70% (12.32%, 15.17%)		7.30% (5.79%, 9.05%)		18.97% (16.85%, 21.24%)		11.67% (8.99%, 14.35%)	<0.001
9	11.68% (10.34%, 13.13%)		8.16% (6.47%, 10.12%)		14.41% (12.46%, 16.53%)		6.25% (3.58%, 8.91%)	<0.001
10	12.25% (10.90%, 13.70%)		7.31% (5.65%, 9.28%)		15.39% (13.49%, 17.44%)		8.08% (5.47%, 10.69%)	<0.001
11	13.28% (11.89%, 14.77%)		6.37% (4.84%, 8.22%)		17.72% (15.71%, 19.87%)		11.35% (8.74%, 13.96%)	<0.001
12	12.69% (10.35%, 15.33%)		8.24% (5.24%, 12.21%)		15.28% (12.11%, 18.91%)		7.04% (2.38%, 11.71%)	0.003
Obesity								
Normal	11.33% (10.70%, 11.99%)	<0.001	7.01% (6.24%, 7.84%)	<0.001	14.59% (13.65%, 15.56%)	<0.001	7.57% (6.35%, 8.80%)	<0.001
Overweight	15.50% (14.24%, 16.83%)		10.16% (8.63%, 11.85%)		20.04% (18.14%, 22.04%)		9.88% (7.39%, 12.37%)	<0.001
Obesity	29.40% (26.73%, 32.17%)		21.23% (17.84%, 24.93%)		37.06% (33.09%, 41.17%)		15.83% (10.58%, 21.09%)	<0.001

Socioeconomic index								
Father's education level, y[a]								
~9	14.22% (13.40%, 15.07%)	0.256	8.36% (7.23%, 9.59%)	0.098	16.97% (15.91%, 18.08%)	0.465	8.62% (7.04%, 10.20%)	<0.001
~12	13.06% (12.13%, 14.04%)		8.97% (7.93%, 10.09%)		18.42% (16.79%, 20.13%)		9.45% (7.49%, 11.41%)	<0.001
~15	13.81% (12.04%, 15.74%)		10.73% (8.61%, 13.16%)		17.48% (14.60%, 20.66%)		6.75% (3.06%, 10.44%)	<0.001
>15	17.05% (9.87%, 26.55%)		14.75% (6.98%, 26.17%)		22.22% (8.62%, 42.26%)		7.47% (−10.56%, 25.5%)	0.417
Father's occupation[b]								
Manager	13.29% (11.18%, 15.64%)	0.110	11.31% (8.34%, 14.89%)	0.007	14.71% (11.84%, 17.97%)	0.006	3.39% (−0.94%, 7.73%)	0.125
Worker	14.82% (13.73%, 15.95%)		8.25% (7.00%, 9.65%)		19.80% (18.18%, 21.49%)		11.55% (9.46%, 13.63%)	<0.001
Techni-cist/re-searcher	13.84% (11.11%, 16.94%)		13.42% (10.16%, 17.27%)		14.66% (9.97%, 20.49%)		1.24% (−4.84%, 7.31%)	0.689
Farmer	13.89% (12.90%, 14.92%)		8.24% (6.98%, 9.66%)		17.17% (15.81%, 18.59%)		8.92% (7.02%, 10.82%)	<0.001
Others	12.60% (11.47%, 13.81%)		9.20% (7.84%, 10.72%)		16.12% (14.33%, 18.04%)		6.92% (4.62%, 9.22%)	<0.001
Mother's occupation[c]								
Manager	13.76% (11.11%, 16.77%)	0.096	10.79% (7.17%, 15.41%)	0.104	15.75% (12.15%, 19.91%)	0.268	4.87% (−0.54%, 10.28%)	0.078
Worker	14.32% (13.18%, 15.52%)		8.84% (7.51%, 10.31%)		19.16% (17.40%, 21.01%)		10.24% (8.00%, 12.49%)	<0.001
Techni-cist/re-searcher	16.74% (12.23%, 22.09%)		15.15% (9.51%, 22.43%)		18.69% (11.81%, 27.38%)		3.54% (−6.05%, 13.13%)	0.469
Farmer	12.74% (11.74%, 13.79%)		8.79% (7.61%, 10.09%)		16.73% (15.15%, 18.41%)		8.06% (6.06%, 10.07%)	<0.001
Others	14.26% (13.27%, 15.29%)		8.62% (7.33%, 10.06%)		17.40% (16.06%, 18.79%)		8.78% (6.89%, 10.68%)	<0.001
Income, RMB[d]								

~500	16.26% (13.67%, 19.12%)	0.009	9.78% (4.57%, 17.76%)	0.880	17.18% (14.35%, 20.32%)	0.988	7.40% (0.67%, 14.13%)	0.031
~1000	16.06% (14.17%, 18.09%)		8.51% (5.28%, 12.84%)		17.59% (15.44%, 19.90%)		9.08% (4.89%, 13.26%)	<0.001
~2000	15.13% (13.69%, 16.65%)		8.29% (6.26%, 10.73%)		17.65% (15.86%, 19.54%)		9.35% (6.54%, 12.17%)	<0.001
~3000	14.33% (13.04%, 15.69%)		8.85% (7.20%, 10.73%)		17.72% (15.93%, 19.62%)		8.87% (6.37%, 11.37%)	<0.001
>3000	13.00% (12.04%, 14.01%)		9.50% (8.37%, 10.72%)		17.13% (15.55%, 18.81%)		7.64% (5.66%, 9.62%)	<0.001
Living with grandparents[e]								
No	13.97% (13.29%, 14.66%)	0.039	9.27% (8.39%, 10.21%)	0.355	17.06% (16.12%, 18.03%)	0.099	7.79% (6.48%, 9.09%)	<0.001
Yes	15.81% (14.16%, 17.57%)		8.03% (5.85%, 10.70%)		18.97% (16.86%, 21.23%)		10.94% (7.78%, 14.11%)	<0.001
People live with child[f]								
1	15.44% (12.60%, 18.63%)	0.106	9.09% (5.06%, 14.78%)	0.687	17.72% (14.22%, 21.66%)	0.802	8.62% (2.82%, 14.43%)	0.004
2~3	13.58% (12.73%, 14.46%)		8.82% (7.76%, 9.98%)		17.08% (15.85%, 18.36%)		8.25% (6.60%, 9.91%)	<0.001
4	14.86% (13.89%, 15.88%)		9.60% (8.25%, 11.08%)		17.67% (16.37%, 19.02%)		8.07% (6.16%, 9.97%)	<0.001
Medical insurance[g]								
No	13.83% (13.16%, 14.52%)	0.004	9.26% (8.39%, 10.19%)	0.209	17.01% (16.06%, 17.99%)	0.096	11.59% (8.28%, 14.91%)	<0.001
Yes	16.41% (14.72%, 18.20%)		7.31% (4.91%, 10.39%)		18.88% (16.87%, 21.02%)		7.79% (6.49%, 9.10%)	<0.001
Perinatal measures								
Gestational hypertension[h]								
No	14.02% (13.38%, 14.67%)	0.024	8.96% (8.12%, 9.84%)	0.018	17.21% (16.33%, 18.12%)	0.297	8.25% (7.02%, 9.48%)	<0.001

Yes	20.00% (14.34%, 26.70%)		18.18% (9.08%, 30.90%)		20.83% (13.96%, 29.20%)		2.65% (−9.87%, 15.17%)	0.678
Birth weight, g[i]								
~3000	15.59% (14.35%, 16.90%)	0.026	9.89% (8.21%, 11.77%)	0.110	18.75% (17.08%, 20.51%)	0.149	8.86% (6.44%, 11.29%)	<0.001
~3600	14.04% (13.09%, 15.03%)		9.52% (8.29%, 10.87%)		17.14% (15.81%, 18.54%)		7.62% (5.76%, 9.47%)	<0.001
>3600	13.35% (12.26%, 14.50%)		7.70% (6.31%, 9.28%)		16.58% (15.08%, 18.16%)		8.88% (6.79%, 10.98%)	<0.001
Breastfeeding, month[j]								
0~3	13.74% (12.50%, 15.06%)	<0.001	10.42% (8.88%, 12.13%)	0.090	17.05% (15.13%, 19.11%)	0.084	6.63% (4.11%, 9.15%)	<0.001
4~10	13.21% (12.33%, 14.13%)		8.28% (7.16%, 9.50%)		16.51% (15.26%, 17.81%)		8.23% (6.53%, 9.93%)	<0.001
>10	16.31% (15.06%, 17.63%)		9.20% (7.32%, 11.38%)		18.73% (17.20%, 20.33%)		9.53% (7.02%, 12.03%)	<0.001
Father with obesity[k]								
No	14.02% (13.36%, 14.70%)	0.149	8.99% (8.12%, 9.91%)	0.126	17.14% (16.23%, 18.08%)	0.263	8.42% (7.10%, 9.73%)	<0.001
Yes	16.19% (14.17%, 18.38%)		10.00% (7.41%, 13.11%)		19.92% (17.14%, 22.93%)		8.05% (4.91%, 11.20%)	<0.001
Mother with obesity[l]								
No	14.03% (13.35%, 14.73%)	0.040	8.78% (7.88%, 9.73%)	0.474	17.19% (16.26%, 18.16%)	0.055	8.15% (6.88%, 9.43%)	<0.001
Yes	15.29% (13.70%, 16.99%)		10.52% (8.44%, 12.90%)		18.57% (16.33%, 20.98%)		9.92% (5.98%, 13.86%)	<0.001
Physical activity								
Weekday, min/day[m]								
~10	14.89% (14.02%, 15.79%)	0.028	9.26% (8.09%, 10.54%)	0.637	17.99% (16.82%, 19.20%)	0.141	8.73% (7.05%, 10.41%)	<0.001
>10	13.47% (12.58%, 14.40%)		8.86% (7.72%, 10.11%)		16.68% (15.41%, 18.00%)		7.82% (6.08%, 9.56%)	<0.001
Weekend, min/day[n]								

~55	14.37% (13.59%, 15.17%)	0.560	8.91% (7.88%, 10.03%)	0.590	17.43% (16.38%, 18.51%)	0.900	8.52% (7.03%, 10.01%)	<0.001
>55	13.97% (12.92%, 15.08%)		9.38% (8.04%, 10.87%)		17.31% (15.80%, 18.90%)		7.92% (5.86%, 9.99%)	<0.001
Total, min/week°								
~150	14.83% (13.96%, 15.74%)	0.050	9.17% (7.99%, 10.46%)	0.808	17.87% (16.70%, 19.09%)	0.247	8.70% (7.01%, 10.39%)	<0.001
>150	13.57% (12.69%, 14.50%)		8.96% (7.83%, 10.20%)		16.84% (15.57%, 18.17%)		7.88% (6.15%, 9.61%)	<0.001

Early work in 1972 outlined the linear association between hypertension and serum uric acid with elevated levels of the latter serving as an independent risk factor for hypertension. The level of uric acid was more than 6.5 mg/dl in 25–40% of adults with hypertension. Further, the risk of hypertension was more when the uric acid levels were more than 7 mg/dl. 90% of adolescents who were diagnosed with hypertension (recent diagnosis) manifested higher uric acid levels. Interestingly, lowering the level of uric acid employing drugs further lowered the BP in adolescents (Orlando, Cazzaniga, Giussani, Palestini and Genovesi, 2018).

Further, the elevated uric acid is thought to inhibit nitric oxide release from endothelial cells while activating the renin–angiotensin system to result in renal vasoconstriction and hence, high blood pressure.

Figure 5 below is illustrative of the plausible mechanism of elevated uric acid to cause high blood pressure (Copyright: © 2018 Orlando, Cazzaniga, Giussani, Palestini and Genovesi. *This is an open-access article distributed under the terms of the* ***Creative Commons Attribution License (CC BY)****. The use, distribution or reproduction in other forums is permitted, provided the original author(s) and the copyright owner are credited and that the original publication in this journal is cited, in accordance with accepted*

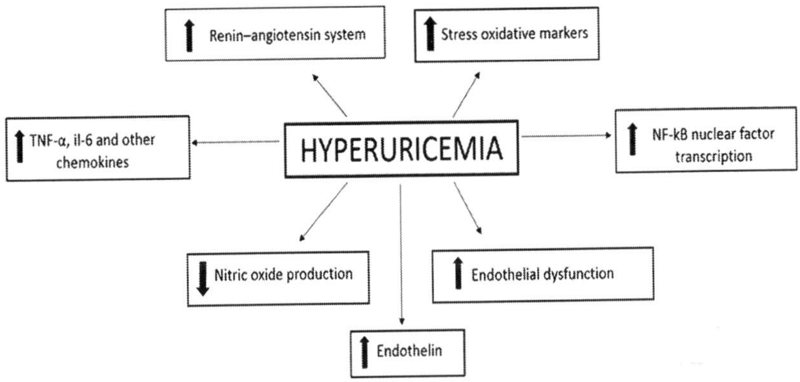

A systematic review and meta-analysis in the *Journal of Translational Medicine* of 14 studies with 93873 participants published in 2020 scrutinized the impact of sugar-sweetened beverages (SSBs) on the blood pressure of children and adolescents. Analyses revealed that the systolic blood pressure increased by 1.67 mmHg with high consumption of SSBs. The probability to develop hypertension was increased in those who consumed high amounts of SSBs vs. low consumers by a factor of 1.36(Farhangiet al, 2020).

This becomes relevant given the trend of consuming such foods; 8.4% of the daily energy intake is accounted for by SSBs in 2–19-year-olds as per the US Nutrition Examination Survey (Kit et al, 2013)! Perhaps, it is time to look at the diet to quell such disorders.

Table 16 below outlines the studies assessing the consumption of SSB and hypertension in children and adolescents(Farhangiet al, 2020:

First author	**Country**	**Number**	**Age range (y)**	**SSB dose (mean/median g/d)**	**SSB type**	**Main Results**
Ambrosini GL [8]	Australia	1433/478	14, 17	47.5±37.1	Carbonated+cordials or and non-100% fruit juice/	SBP in higher tertiles of SSB intake was higher than lower tertile (P=0.03). No difference in DBP was found.
BarstadLH [37]	Norway	313/(62-70)	12–18	None to at least 4 glasses per week (~ 1375)	Sugar-sweetened soda	SBP in higher intakes of SSB was higher than lower.
Bortsov AV [26]	USA	902/ (304-600)	10–22	0-750	Sugar-sweetened soda	No significant difference in SBP or DBP between three categories SSB intake
Bremer AA [39]	USA	2630/ (876)	12-19	Low (≤20th percentile) to high (≥80th percentile) of the sum of the number of SSB serving equivalents	Caloric soft drinks, colas, sugar-sweetened fruit drinks, or other SSBs	Significantly higher SBP values (P=0.03) and no difference in DBP values between low and high SSB consumers
Chan TF [13]	Taiwan	2727/ (242-196)	12–16	Non to>750	Any type	No significant difference between SBP and DBP of different SSB categories

Chan TF [13]	Taiwan	2727/ (406-120)	12–16	Non to > 750	Any type	Significantly higher SBP in higher intakes compared with lower intakes of SSB (P=0.043). No difference in DBP was observed.
DeBoer EC [19]	USA	9600	5–6	0-750	Chocolate milk, yogurt drinks, lemonades, juices and soft drinks	Significantly higher SBP in higher versus lower SSBtertiles
DeBoerEC [19]	USA	2516/ (794-905)	11-12	0-950	Yogurt drinks, soft drinks, juices, lemonades, sport drinks and energy drinks	Significantly higher SBP in higher versus lower SSBtertiles
GuiZH [23]	China	53,151/ (15 763-17773)	6-17	0-500	Coca-Cola, Sprite, orange juice, Nutrition Express, and Red Bull	Significantly higher SBP and DBP in higher versus lower SSB intakes (P<0.001), no difference in odds of HTN in different SSB intakes.

He B [24]	China	2032/ (440-705)	6–18	0-120	Carbonated drinks, juices, and sports and sweet tea beverages.	Significantly higher SBP and DBP in higher versus lower SSB intakes (P<0.001).
Lin WT [38]	Taiwan	2727/ (164-317)	12-16	non-intake to≥1000 ml/d	SSB, including soft drinks, fruit drinks and sweetened teas.	Significant increase in SBP (3.47 mmHg; P=0.004) and no significant change in DBP (p=0.514) in higher versus lower SSB consumers.
Loh DA [27]	Malaysia	881/ (293)	13	338.75	Carbonate beverages	No significant difference in SBP and DBP between SSBtertiels.
Mirmiran et al. [15]	Iran	424/ (106)	6–18	132.7	Sugar sweetened carbonated soft drinks (SSSDs) and fruit juice drinks (non-100% fruit juices)	Significantly higher SBP in highest versus lowest SSB category (P=0.021). No difference in DBP between SSB quartiles (P=0.52). Higher odds of HTN in highest versus lowest SSB category (2.90 (0.91–9.26); P=0.043)
Qin Z [9]	China	10091/ (249-203)	Grade 4: 9.04 ± 0.38 Grade 7: 12.03 ± 0.41	Consumers/ non-consumers	Sprite and Coca-Cola	Higher odds of HTN in SSB consumers versus non-consumers [OR:1.40 (1.15,1.70)]

Souza BSN [10]	Brazil	488/ (419-25)	9–16	500	Soft drinks, fruit drinks and sweet-ened teas	Significantly higher SBP and DBP in SSB consumers than non-consumers (P < 0.05)
Zhu Z [11]	China	3958/ (343-2582)	6-17	201.7	Nonal-coholic bever-ages sweet-ened by sugar, exclud-ing fresh juice.	Significantly high-er SBP and DBP in high consumers versus low con-sumers (P<0.001; P=0.004)

The plausible mechanism has been succinctly put forth by Orlando and team (2018) with the consumption of fructose accounting for an increase in uric acid in half an hour. Following the uptake of fructose by GLUT5 in the intestines and its transport, fructose-1-phosphate is produced by the action of fructokinase in the liver. This results in the formation of fructose-1-phosphate to dampen the levels of ATP in the cells with the formation of uric acid from the AMP formed.

Figure 6 below highlights the increase in uric acid that is a factor for hypertension following fructose consumption raising the importance of looking at the diet in children (Copyright: © 2018 Orlando, Cazzaniga, Giussani, Palestini and Genovesi. *This is an open-access article distributed under the terms of the* ***Creative Commons Attribution License (CC BY)****. The use, distribution or reproduction in other forums is permitted, provided the original author(s) and the copyright owner are credited and that the original publication in this journal is cited, in accordance with accepted academic practice. No use, distribution or reproduction is permitted which does not comply with these terms*):

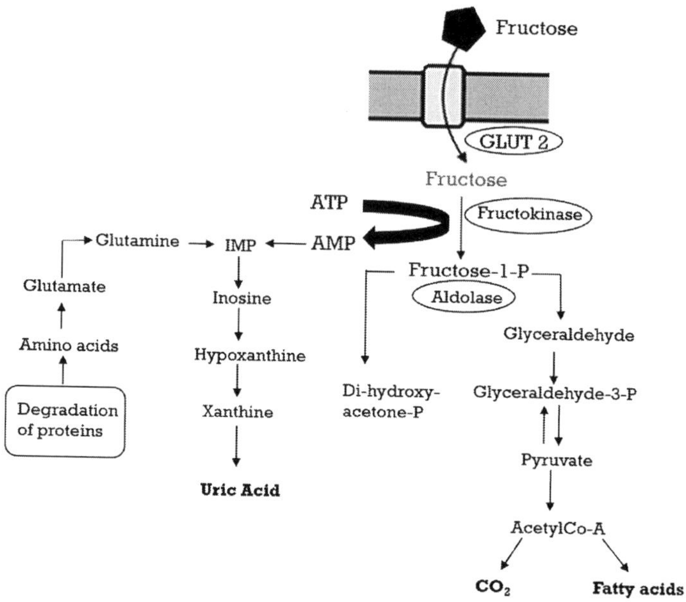

2.5 A NEONATAL LINK

The other "side" of the end product of heme catabolism, bilirubin has been outlined by population-based research where its negative association with stroke, cardiovascular disease and hypertension has been presented. Given that the neonatal stage is linked with high bilirubin levels in the serum, a secondary data analysis-based study entailed the assessment of the role of high levels of neonatal serum bilirubin levels and childhood hypertension. On the basis of the Collaborative Perinatal Project (CPP) having 46,021 women with 56,990 pregnancies, screening yielded 37,544 infants: 31,819 term infants and 5,725 preterm infants. While the blood pressure displayed no significant association with total serum bilirubin (TSB), conjugated bilirubin (CB) and unconjugated bilirubin (UCB) levels, a finding emerged. Low levels of TSB were linked to a higher systolic blood pressure of 0.62 mmHg while high UCB was associated with an increase in systolic blood pressure by 0.70 mmHg in preterm infants. This could not be seen in term infants at the age of seven. This involvement of UCB in hypertension opens up the avenue for further research where this molecule could emerge as a vital player in managing the condition (Yu et al, 2019).

Table 17 below outlines the profile of neonatal serum bilirubin and the blood pressure: ***TOP***: The bilirubin levels 48-h post-birth (© 2019 Yu et al. *This is an open access article distributed under the terms of the Creative*

Commons Attribution License, which permits unrestricted use, distribution, and reproduction in any medium, provided the original author and source are credited):

β (95% CI)		**Systolic Blood Pressure at 7 years old**		**Diastolic Blood Pressure at 7 years old**	
		P	**β (95% CI)**	***P***	
Total subjects	Total Serum Bilirubin	0.23 (-0.01, 0.47)	0.06	-0.10 (-0.21, 0.02)	0.08
	Conjugated Bilirubin	0.24 (-0.02, 0.50)	0.08	-0.08 (-0.33, 0.17)	0.51
	Unconjugated Bilirubin	0.00 (-0.11, 0.11)	0.99	0.03 (-0.08, 0.14)	0.55
Preterm infants	Total Serum Bilirubin	0.62 (0.38, 0.86)	<0.001	-0.03 (-0.27, 0.21)	0.79
	Conjugated Bilirubin	0.44 (-0.22, 1.10)	0.19	-0.16 (-0.49, 0.17)	0.62
	Unconjugated Bilirubin	0.70 (0.10, 1.30)	<0.05	-0.09 (-0.39, 0.21)	0.56
Term infants	Total Serum Bilirubin	-0.05 (-0.23, 0.08)	0.47	-0.13 (-0.27, 0.01)	0.06
	Conjugated Bilirubin	0.25 (-0.02, 0.52)	0.09	-0.07 (-0.33, 0.19)	0.58
	Unconjugated Bilirubin	-0.08 (-0.20, 0.04)	0.18	0.01 (-0.11, 0.13)	0.87

BOTTOM: Subtypes of bilirubin and blood pressure (© 2019 Yu et al. *This is an open access article distributed under the terms of the Creative Commons Attribution License, which permits unrestricted use, distribution, and reproduction in any medium, provided the original author and source are credited*):

Bilirubin n/N (%)		**High blood pressure at the age of 7 years**						
		Model 1		**Model 2**		**Model 3**		
		OR	**95% CI**	**OR**	**95% CI**	**OR**	**95% CI**	
Uncon-jugated bilirubin [a]	< 3mg/dl	29/317 (9.2)	1	-	1	-	1	-
	≥ 3mg/dl, < 6mg/dl	43/326 (13.2)	1.51	0.92, 2.49	1.54	0.93, 2.55	1.55	0.93, 2.57
	≥ 6mg/dl, <9mg/dl	89/659 (13.5)	1.61	1.04, 2.50	1.55	0.99, 2.43	1.55	1.00, 2.46
	≥ 9mg/dl, < 12mg/dl	75/539 (13.9)	1.63	1.04, 2.56	1.51	0.95, 2.41	1.51	0.95, 2.42
	≥ 12mg/dl	109/728 (15.0)	1.79	1.16, 2.75	1.54	0.97, 2.44	1.54	0.96, 2.46
Conjugated bilirubin[b]	< 1mg/dl	272/2074 (13.1)	1	-	1	-	1	-
	≥ 1mg/dl, < 2mg/dl	58/378 (15.4)	1.20	0.88, 1.63	1.11	0.80, 1.53	1.08	0.73, 1.53
	≥ 2mg/dl	15/117 (12.8)	0.95	0.55, 1.66	0.85	0.48, 1.50	0.88	0.47, 1.58

Model 1: crude odds ratios; Model 2: Adjusted for race (white, black, and other races), sex (male and female), gestational age (as a categorical variable), birth weight (<2500g, and ≥2500g), transfusion (yes and no), hypertensive disorders during pregnancy (none, moderate, and severe), maternal smoking (0, 1–19 and 20 cigarette per day during pregnancy) and socioeconomic status (comprised of 5 categories as assessed by the original CPP investigators); Model 3: adjusted for the same factors as model 2 in Generalized Estimating Equation model

[a] Adjusted by direct serum bilirubin additionally;

[b] Adjusted by indirect serum bilirubin additionally;

2.6 HYPERTENSION AND SLEEP

Research has pointed out the importance of optimal sleep for health with studies highlighting the vital role of sleep deprivation on hypertension in adults. While the involvement of sleep in growth and maturation in adolescents is established, research points out sleep issues among adolescence (Kuciene et al, 2014). An example is a meta-analysis of 41 surveys of sleep patterns across the globe for 1999-2010: increasing age was associated with sleep patterns with later bedtimes observed in Asian adolescents vs. their North American and European counterparts. Further, a delay in sleep patterns was observed with age increase (Gradisar et al, 2011). In this section, let us see few studies that outlined the association between hypertension and sleep.

Kuciene et al (2014) explored the patterns of sleep deprivation and hypertension in Kaunas city and in Kaunas district (Lithuania: known to have a high prevalence of hypertension). The cross-sectional study (***reported as the first***) was inclusive of 6,940 participants aged 12 to 15 years. 12.6% of the adolescents showed prehypertension while 22.5% displayed hypertension. Boys demonstrated a higher extent of hypertension vs. girls: 29.6% vs. 16.4%. Of the adolescents who reported a sleep duration of <7 h/day (8.7% of the participants), 31.5% had hypertension while this condition was seen in 29.2% of participants whose sleep was 7–<8 h/day (21.0%). Subsequent analyses revealed a significant association of prehypertension and hypertension for both sexes and insufficient sleep (7–<8 h/day and <7 h/day) vs. those who slept ≥8 h/day.

Table 18 below depicts ***TOP***: the trend in blood pressure across the adolescents: here adolescents with prehypertension and hypertension demonstrated risk factors such as inadequate sleep, smoking, obesity and less physical activity (Kuciene et al, 2014: *This article is published under license to BioMed Central Ltd. This is an Open Access article distributed under the terms of the Creative Commons Attribution License (http://creativecommons.org/licenses/by/2.0), which permits unrestricted use, distribution, and reproduction in any medium, provided the original work is properly credited. The Creative Commons Public Domain Dedication waiver (http://creativecommons.org/publicdomain/zero/1.0/) applies to the data made available in this article, unless otherwise stated*):

Key:

¶*Values are numbers and percentages.*

#Values are means ± SD (standard deviations).

BMI body mass index, *SBP* systolic blood pressure, *DBP* diastolic blood pressure.

**Significant differences between the groups were determined by the chi-squared (χ^2) test.*

***Significant differences between the three groups were determined by Kruskal-Wallis test and ANOVA (data in italics).*

[a]*Significantly different ($P < 0.05$) from normotensive participants (Mann–Whitney U* test; *t-test).*

[b]*Significantly different ($P < 0.05$) from prehypertensive participants (Mann–Whitney U* test; *t-test).*

[a,b]*Superscript letters in bold and italics indicate that significant differences ($P < 0.05$) between groups were determined by the t-test.*

Variables	**Normo-tensive (n=4504)**		**Prehyperten-sive (n=875)**		**Hypertensive (n=1561)**		**P***
Categorical variables¶	**N**	**%**	**n**	**%**	**n**	**%**	
Sex:							
Boys	1,770	39.3	475	54.3	944	60.5	<0.001
Girls	2,734	60.7	400	45.7	617	39.5	
Age (years):							
12–13	2,608	57.9	292	33.4	694	44.5	<0.001
14–15	1,896	42.1	583	66.6	867	55.5	
Sleep duration (h/day):							
≥8	3,406	75.6	529	60.5	947	60.6	<0.001
7– <8	796	17.7	234	26.7	424	27.2	
<7	302	6.7	112	12.8	190	12.2	
BMI categories:							
Normal weight	4,108	91.2	691	79.0	1,140	73.0	<0.001
Overweight	349	7.8	149	17.0	337	21.6	
Obesity	47	1.0	35	4.0	84	5.4	
Physical activity (h/day):							
≥1	4,066	90.3	762	87.1	1,351	86.5	<0.001

0.5– <1	339	7.5	85	9.7	169	10.8	
<0.5	99	2.2	28	3.2	41	2.7	
Smoking:							
No	4,171	92.6	782	89.4	1,444	92.5	0.003
Yes	333	7.4	93	10.6	117	7.5	
Continuous variables [#]	**mean±SD**		**mean±SD**		**mean±SD**		**P****
Age (years)	13.29±1.09		13.89±1.03[a]		13.62±1.09[a,b]		<0.001
Weight (kg)	49.84±10.15		59.54±11.31[a]		60.22±12.66[a]		<0.001
Height (cm)	162.37±9.32		169.75±8.87[a]		168.20±9.90[a,b]		<0.001
BMI (kg/m^2)	18.77±2.76		20.59±3.22[a]		21.17±3.50[a,b]		<0.001
SBP (mmHg)	109.72±6.93		124.85±3.90[a]		139.41±9.85[a,b]		<0.001
DBP (mmHg)	63.10±6.38		67.22±6.91[a]		71.85±8.08[a,b]		*<0.001*

BOTTOM: The trends in hypertension across both sexes with insufficient sleep (Kuciene et al, 2014: *This article is published under license to BioMed Central Ltd. This is an Open Access article distributed under the terms of the Creative Commons Attribution License (http://creativecommons.org/licenses/by/2.0), which permits unrestricted use, distribution, and reproduction in any medium, provided the original work is properly credited. The Creative Commons Public Domain Dedication waiver (http://creativecommons.org/publicdomain/zero/1.0/) applies to the data made available in this article, unless otherwise stated*):

Key: OR – crude odds ratio; aOR[1] – adjusted odds ratios for age, body mass index, physical activity, and smoking; CI – confidence interval. *All results were significant at P<0.001, except (* – P<0.05)*

Vari-ables	**Normal BP**		**Prehypertension**				**Hypertension**			
	n	**%**	**n**	**%**	**OR (95% CI)**	**aOR[1] (95% CI)**	**n**	**%**	**OR (95% CI)**	**aOR[1] (95% CI)**
Boys:										
Sleep duration (h/day):										
≥8	1424	80.5	316	66.5	1.00	1.00	612	64.8	1.00	1.00

7–<8	229	12.9	112	23.6	2.20 (1.71–2.85)	1.86 (1.41–2.44)	235	24.9	2.39 (1.95–2.93)	2.13 (1.72–2.65)
<7	117	6.6	47	9.9	1.81* (1.26–2.59)	1.49* (1.01–2.19)	97	10.3	1.93 (1.45–2.57)	1.70* (1.26–2.31)
Girls:										
Sleep duration (h/day):										
≥8	1982	72.5	213	53.3	1.00	1.00	335	54.3	1.00	1.00
7–<8	567	20.7	122	30.4	2.00 (1.57–2.55)	1.81 (1.41–2.32)	189	30.6	1.97 (1.61–2.41)	2.00 (1.62–2.46)
<7	185	6.8	65	16.3	3.27 (2.38–4.48)	3.01 (2.17–4.18)	93	15.1	2.97 (2.26–3.92)	3.07 (2.31–4.09)

Another study published in 2018 in the *Journal of clinical research in pediatric endocrinology* was elucidated as "*the first to demonstrate a relationship between short sleep duration and high BP in children and adolescents aged 11-17 years*" (Bal et al, 2018). The study was inclusive of 2860 adolescents (49.6% girls) in primary and secondary schools in Kayseri in Turkey. 35.0% and 30.8% of boys and girls, respectively demonstrated prehypertension and hypertension. A lowering in these conditions was observed with an increase in the sleep duration. Further, an increase in sleep time resulted in a dip in mean SBP from 112.9 mmHg to 107.9 mmHg in girls and a drop in mean DBP from 72.7 mmHg to 69.4 mmHg in boys.

Table 19 below documents ***TOP***: sleep time, BP and risk factors (Bal et al, 2018: © Copyright 2018, Journal of Clinical Research in Pediatric Endocrinology, Published by Galenos Publishing. *This is an open-access article distributed under the terms of the Creative Commons Attribution License, which permits unrestricted use, distribution, and reproduction in any medium, provided the original work is properly cited*):

Variables	Boys (n = 1385) Sleep duration categories (hours)					Girls (n = 1475) Sleep duration categories (hours)				
	≤8 (n = 750)	8.1-8.9 (n = 360)	9.0-9.9 (n = 205)	≥10 (n = 70)	p	≤8 (n = 844)	8.1-8.9 (n = 335)	9.0-9.9 (n = 222)	≥10 (n = 74)	p
BMI	20.7 ± 0.1	20.4 ± 0.2	20.1 ± 0.2	20.4 ± 0.4	0.073	20.9 ± 0.1	20.7 ± 0.2	20.6 ± 0.2	20.2 ± 0.4	0.177
WC	70.4 ± 0.4	69.8 ± 0.4	69.9 ± 0.6	69.7 ± 0.9	0.714	65.7 ± 0.2	65.2 ± 0.4	65.1 ± 0.4	64.3 ± 0.8	0.206
WHtR	0.43 ± 0.003	0.42 ± 0.003	0.43 ± 0.003	0.43 ± 0.006	0.936	0.42 ± 0.001	0.41 ± 0.002	0.42 ± 0.003	0.41 ± 0.005	0.376
MUAC	22.3 ± 0.1	22.0 ± 0.2	22.0 ± 0.2	21.9 ± 0.4	0.223	22.1 ± 0.1	22.2 ± 0.1	22.1 ± 0.2	22.0 ± 0.3	0.973
TSF	9.2 ± 0.3	9.7 ± 0.2	9.7 ± 0.3	9.2 ± 0.6	0.641	14.3 ± 0.2	14.3 ± 0.3	14.1 ± 0.4	13.6 ± 0.6	0.752
AFA	10.3 ± 0.2	10.1 ± 0.3	9.7 ± 0.4	9.6 ± 0.4	0.562	14.4 ± 0.2	14.4 ± 0.5	14.2 ± 0.4	13.6 ± 0.7	0.790
SBP	114.9 ± 0.5	114.5 ± 0.7	113.9 ± 0.9	112.5 ± 1.5	0.436	**112.9 ± 0.5**^{a}	**112.1 ± 0.7**^{ab}	**111.5 ± 0.9**^{ab}	**107.9 ± 1.5**^{b}	**0.013**
DBP	**72.7 ± 0.4**^{a}	**71.0 ± 0.5**^{ab}	**71.0 ± 0.7**^{ab}	**69.4 ± 1.2**^{b}	**0.007**	71.5 ± 0.3	71.7 ± 0.5	71.4 ± 0.7	69.1 ± 1.1	0.235
Fat %	24.7 ± 0.3	25.2 ± 0.5	24.3 ± 0.7	24.3 ± 1.1	0.698	35.6 ± 0.4	35.6 ± 0.6	35.1 ± 0.7	34.4 ± 1.2	0.751

Values are expressed as age-adjusted means (standard error of the mean), in accordance with where age or age and body mass index were the covariates for multiple comparisons (Bonferroni). Statistical significancy was ($p < 0.05$)

BMI: body mass index, WHtR: waist/height ratio, WC: waist circumference, MUAC: mid-upper arm circumference, TSF: triceps skinfold thickness, AFA: arm fat area, SBP: Systolic blood pressure, DBP: diastolic blood pressure, Fat %: arm fat area

$^{a, b}$: According to post-hoc tests, statistically significant ($p < 0.05$) different anthropometric indices between sleep duration categories were labelled with different letters

BOTTOM: Analyses of prehypertension and its association with sociodemographic and behavioral factors in the study (Bal et al, 2018: © Copyright 2018, Journal of Clinical Research in Pediatric Endocrinology, Published by Galenos Publishing. *This is an open-access article distributed under the terms of the Creative Commons Attribution License, which permits* *unrestricted use, distribution, and reproduction in any medium, provided the original work is properly cited*):

Variables	Gender					
	Boys		Girls		Boys	Girls
	Normotensive	Prehypertensive	Normotensive	Prehypertensive	Univariate logistic regression (Adjusted for age)	
Continuous Variables	$\bar{X}$ + SD	$\bar{X}$ + SD	$\bar{X}$ + SD	$\bar{X}$ + SD	OR (CI)	OR (CI)
Computer use (h)	0.75 ± 0.05	0.83 ± 0.06	0.43 ± 0.03	0.46 ± 0.06	1.04 (0.97-1.13)	1.02 (0.93-1.12)
Sleep duration (h)	**8.48 ± 0.04**	**8.31 ± 0.06**	**8.47 ± 0.04**	**8.27 ± 0.06**	**0.89 (0.82-0.98)**	**0.88 (0.81-0.97)**
Viewing TV (h)	2.94 ± 0.05	2.96 ± 0.07	2.70 ± 0.00	2.75 ± 0.07	1.01 (0.94-1.08)	1.02 (0.95-1.10)
BMI (kg/m²)	**20.13 ± 0.09**	**21.20 ± 0.15**	**20.50 ± 0.09**	**21.44 ± 0.16**	**1.12 (1.08-1.16)**	**1.09 (1.06-1.14)**
Categorical Variables	n (%)	n (%)	n (%)	n (%)		
Sleep duration categories (h)						
≤8.0	**458 (61.1)**	**292 (38.9)**	**566 (67.1)**	**278 (32.9)**	1.0	1.0
8.1-8.9	**250 (69.4)**	**110 (30.6)**	**238 (71.0)**	**97 (29.0)**	**0.69 (0.53-0.90)**	0.83 (0.63-1.09)
9.0-9.9	**141 (68.8)**	**64 (31.2)**	**155 (69.8)**	**67 (30.2)**	**0.71 (0.51-0.99)**	0.88 (0.64-1.21)
≥10.0	**51 (72.9)**	**19 (27.1)**	**61 (82.4)**	**13 (17.6)**	**0.58 (0.34-0.99)**	**0.43 (0.23-0.80)**
Location of the school						
Suburban	**601 (62.5)**	**360 (37.5)**	**646 (65.5)**	**341 (34.5)**	**1.0**	**1.0**
Urban	**299 (70.5)**	**125 (29.5)**	**374 (76.6)**	**114 (23.4)**	**1.43 (1.12-1.83)**	**1.73 (1.35-2.22)**
Transportation to school						
On foot	582 (66.1)	299 (33.9)	657 (69.8)	284 (30.2)	1.0	1.0
By bus/private car	318 (63.1)	186 (36.9)	363 (68.0)	171 (32.0)	1.14 (0.91-1.43)	1.09 (0.87-1.37)
Household income						
Poor	**174 (72.2)**	**67 (27.8)**	140 (71.4)	56 (28.6)	**1.0**	1.0
Fair	**512 (64.3)**	**284 (35.7)**	549 (67.9)	259 (32.1)	**1.63 (1.14-2.32)**	1.06 (0.73-1.53)
Good	**214 (61.5)**	**134 (38.5)**	331 (70.3)	140 (29.7)	**1.44 (1.05-1.99)**	1.18 (0.84-1.66)
House size (m²)						
< 100	262 (66.2)	134 (33.8)	291 (70.5)	122 (29.5)	1.0	1.0
100-200	591 (10.4)	329 (35.8)	683 (68.6)	312 (31.4)	0.92 (0.53-1.58)	1.09 (0.62-1.90)
> 200	47 (68.1)	22 (31.9)	46 (68.7)	21 (31.3)	1.88 (1.18-3.54)	1.09 (0.85-1.39)
Parental schooling (years)						
Father						
< 5	423 (65.9)	219 (34.1)	440 (70.2)	187 (29.8)	1.0	1.0
6-8	208 (65.8)	108 (34.2)	241 (69.5)	107 (30.7)	1.37 (0.92-2.03)	1.06 (0.72-1.56)
9-11	197 (64.8)	107 (35.2)	235 (67.3)	114 (32.7)	1.05 (0.79-1.40)	1.14 (0.86-1.51)
> 12	72 (58.5)	51 (41.5)	104 (68.9)	47 (31.1)	1.01 (0.76-1.33)	1.05 (0.79-1.40)
Mother						
< 5	715 (65.8)	371 (34.2)	814 (68.8)	370 (31.2)	1.0	1.0
6-8	101 (60.5)	66 (39.5)	106 (70.2)	45 (29.8)	0.64 (0.20-2.00)	0.35 (0.10-1.18)
9-11	72 (62.1)	44 (37.9)	81 (68.6)	37 (31.4)	1.18 (0.79-1.75)	1.01 (0.67-1.51)
> 12	12 (75.0)	4 (25.0)	19 (86.4)	3 (13.6)	1.26 (0.90-1.76)	0.93 (0.65-1.35)
Maternal employment						
Unemployed (Housewife)	858 (65.1)	459 (34.9)	975 (68.8)	442 (31.2)	1.0	1.0
Employed (Has a regular job)	42 (61.8)	26 (38.2)	45 (77.6)	13 (22.4)	1.16 (0.70-1.19)	0.64 (0.34-1.19)
Elevator use						
None	664 (66.4)	336 (33.6)	737 (70.7)	306 (29.3)	1.0	1.0
Habitual	236 (61.3)	149 (38.7)	283 (65.5)	149 (34.5)	1.25 (0.98-1.59)	1.27 (0.99-1.61)

Values represent odds ratio 95% confidence interval adjusted for age or age and body mass index

BMI: body mass index, SD: standard deviation, OR: odds ratio, CI: confidence interval

In a nutshell, the risk of prehypertension and hypertension was lowered by 0.89 and 0.88, respectively with an hourly incremental rise in the sleep duration in the adolescents. For both the genders, analyses revealed sleep duration along with the school location as crucial risk factors with household income emerging as a vital risk factor in boys (Bal et al, 2018).

A recent longitudinal survey described as "the first to observe time trend of sleep duration and BP level among Chinese children aged 7–13 years" published in the *International Journal of Hypertension* scrutinized the association of short sleep time and hypertension in 1000 children of the China Health and Nutrition Survey (2004–2015). The sleep duration was found to dip with increasing age with boys showing a higher systolic BP (SBP) and diastolic BP (DBP) of 96.70 mmHg and 64.00 mmHg, respectively vs. that of girls at 95.32 mmHg and 62.80 mmHg. 15.8% of those who slept <9 hours per day demonstrated elevated BP vs. 10.5% of those who slept ≥9 hours (Lili Huang et al, 2020).

Table 20 below depicts how the BP changes with sleep time (Copyright © 2020 Lili Huang et al. *This is an open access article distributed under the Creative Commons Attribution License, which permits unrestricted use, distribution, and reproduction in any medium, provided the original work is properly cited*):

		SBP			**DBP**			**Elevated BP**		
Over-all	*n*	**Mean (SD)**	*T*	*p*	**Mean (SD)**	*t*	*p*	*N* (%)	χ^2	*p*
<9 h	587	98.83 (13.00)	6.104	<0.001	64.84 (9.29)	4.307	<0.001	98 (16.7)	6.532	0.011
≥9 h	1604	95.05 (12.22)			62.94 (9.09)			200 (12.5)		
Boy										
<9 h	327	99.15 (13.20)	4.149	<0.001	64.63 (9.33)	1.473	0.141	57 (17.4)	1.966	0.161
≥9 h	854	95.77 (12.28)			63.76 (9.07)			121 (14.2)		
Girl										
<9 h	260	98.42 (12.46)	4.715	<0.001	65.10 (9.26)	4.725	<0.001	41 (15.8)	5.056	0.025
≥9 h	750	94.25 (12.10)			62.01 (9.03)			79 (10.5)		

Overall, these studies from several parts of the globe point out the vital involvement of sleep duration on hypertension/prehypertension in adolescents. Further research can throw further light and assist the families to exert control on the sleep duration of children and adolescents.

2.7 THE EFFECT OF SEASONS

A review in *Pediatric Nephrology* by Ziegelasch and team (2020) reported that the systolic BP records a peak in fall and winter with a rise of 0.5-1.5mmHg/°C change in the environmental temperature. Further, the transition from warm to cold seasons was accompanied by a spike in the BP in subjects by 57% in a Greek cohort and by 50% in an Indian cohort.

Table 21 below summarizes the research studies on the variation of blood pressure with seasons in children. *Key: BP blood pressure, DBPv diastolic blood pressure variation, NA not assessed, n (obs.) number of observations, SBPv systolic blood pressure variation. SBPv and DBPv represent the difference of blood pressure mean of the winter minus that of the summer. Some studies only appointed the difference as a slope depending on temperature change as indicated by "/− 1 °C". Elevated BP* was defined as proposed by the "Fourth Report on the Diagnosis, Evaluation, and Treatment of High BP in Children and Adolescents" (Narang et al.) and the "European Society of Hypertension" (Nika et al.)* (Ziegelasch et al ,2020: *Open Access: This article is licensed under a Creative Commons Attribution 4.0 International License, which permits use, sharing, adaptation, distribution and reproduction in any medium or format, as long as you give appropriate credit to the original author(s) and the source, provide a link to the Creative Commons licence, and indicate if changes were made. The images or other third party material in this article are included in the article's Creative Commons licence, unless indicated otherwise in a credit line to the material. If material is not included in the article's Creative Commons licence and your intended use is not permitted by statutory regulation or exceeds the permitted use, you will need to obtain permission directly from the copyright holder. To view a copy of this licence, visit http://creativecommons.org/licenses/by/4.0/*):

	***n* (obs.)**	**Age (years)**	**SBPv (mmHg)**	**DBPv (mmHg)**	**Additional cohort information**
Miersch et al. 2013 [2]	6714	3–21	4.5	2.4	Primarily healthy, Germany
Polat et al. 2006 [6]	547	0–15	5.9	3.6	Primarily healthy, Turkey

Jenner et al. 1987 [7]	1037	9	0.5–0.7/−1 °C	0.5–0.7/−1 °C	Primarily healthy, Australia
Levinson et al. 1985 [8]	4086	5–10	2.8–4.0	no variation	School children, USA
De Swiet et al.1984 [9]	491	4–5	1.5/−1 °C	NA	Primarily healthy, Great Britain
Narang et al. 2018 [10]	14,957	5–15	Elevated BP* 29.4% (winter) vs. 18.7% (summer)		Primarily healthy, India
Nika et al.2019 [11]	2832	6–18	Elevated BP* 5.5% (spring) vs. 2.5% (fall)		School children, Greece

The plausible mechanisms have been described as an inverse correlation between blood pressure and physical activity along with elevated levels of norepinephrine in the urine and plasma in winters. Given the involvement of seasonal variation of blood pressure, it is vital to incorporate this aspect while designing and managing hypertension in children.

2.8 PHYSICAL ACTIVITY

The impact of physical activity [physical activity (PA) and sedentary behavior (SB)] was scrutinized in two cross-sectional studies conducted with adolescents: Healthy Lifestyle in Europe by Nutrition in Adolescence (HELENA) and Brazilian Cardiovascular Adolescent Health (BRACAH). The analyses by de Moraes and team (2013) were inclusive of 3,308 adolescents (12.5–17.5 years old) from the former and 991 adolescents (14.0–17.5 years old) from the latter study. The assessment of PA involved the use of the questionnaire model and accelerometers (over seven days with a minimum of 8 hours recording/day for at least 3 days) while SB was evaluated using questions like "*During weekdays, how many hours do you usually spend watching television*?"-"*During weekdays, how many hours do you usually spend on computers*?-"*During weekdays, how many hours do you usually spend playing video games*?"

It was revealed that the systolic BP increases were associated with low PA with lots of SB in boys. The following of the requisite level of PA (≥60 min/d) quelled the impact of a sedentary lifestyle on the diastolic BP in boys of both HELENA and BRACAH studies while this was observed only for girls of the BRACAH study. Variations in culture, environment and income are thought to account for this difference. Table 22 below is illustrative of ***TOP***: the trend in systolic blood pressure in boys with physical activity and sedentary behavior (© 2013 de Moraes et al. *This is an open-access article distributed under the terms of the Creative Commons Attribution License, which permits unrestricted use, distribution, and reproduction in any medium, provided the original author and source are credited*):

Independents Variables	Null model		Unadjusted		Adjusted*	
	HELENA	BRACAH	HELENA	BRACAH	HELENA	BRACAH
	ß (95% CI)	ß (95% CI)	ß (95% CI)	ß (95% CI)	ß (95% CI)	ß (95% CI)
Fixed Effects Constant	124.81 (124.10–125.51)	119.64 (118.42–120.86)			84.4 (76.1–92.7)	
Physical activity by questionnaire			p=0.005	p=0.331	p=0.066	
≥60 min/d			Ref	Ref	Ref	
<60 min/d			**2.92 (0.90–4.96)**	**1.41 (−1.44–4.27)**	2.08 (−0.14–4.29)	
Sedentary behavior by questionnaire			p=0.435	p=0.811	p=0.361	
<2 h/d			Ref	Ref	Ref	
2–4 h/d			1.18 (−0.77–3.14)	−1.32 (−7.77–5.14)	1.02 (−1.01–3.05)	−0.96 (−7.75–5.84)
>4 h/d			0.37 (−1.59–2.33)	−1.84 (−7.93–4.24)	−0.79 (−2.87–1.30)	−2.04 (−8.42–4.34)
Physical activity by accelerometers **			p=0.399		p=0.762	
≥60 min/d			Ref		Ref	
<60 min/d			0.74 (−0.99–2.47)		−0.30 (−2.21–1.62)	

MPA by accelerometers (min/d)			p=0.895		p=0.859	
		0.13 (−1.78–2.04)		−0.01 (−0.05–0.07)		
VPA by accelerometers (min/d)			p=0.133		p=0.929	
		−0.05 (−0.11–0.01)		−0.01 (−0.06–0.07)		
MPA by accelerometers (min/d)			p=0.646		p=0.936	
		−0.01 (−0.05–0.03)		−0.01 (−0.04–0.04)		
SB by accelerometers (min/d)			p=0.069		p=0.337	
		0.01 (−0.01–0.02)		0.01 (−0.01–0.02)		
Cluster PA and SB by accelerometers *			p=0.673		p=0.465	
<60 min/d+3rd tertile			Ref		Ref	
<60 min/d+2nd tertile			−0.58 (−4.09–2.94)		−2.16 (−5.49–1.16)	
<60 min/d+1st tertile			0.36 (−3.14–3.85)		−0.09 (−3.37–3.18)	
≥60 min/d+3rd tertile			0.88 (−2.56–4.31)		0.58 (−2.65–3.81)	
≥60 min/d+2nd tertile			0.08 (−3.06–3.21)		0.30 (−2.65–3.24)	
≥60 min/d+1st tertile			−0.88 (−3.75–1.98)		−0.06 (−2.86–2.74)	

Intraclass correlation coeficient	0.07	0.06			0.10	0.04
Standard deviation context	3.69	3.2			4.58	2.84
Standard deviation individual	13.98	13.2			13.91	13.40
Akaike Information Criterion	12,191.3	3,560.7			10,520.4	3,347.0

Beta coefficient and their respective confidence intervals 95% (95% CI).

*This analysis was adjusted for potential confounders: *age, socioeconomic status, parental education, regular tobacco smoking, body mass index and waist circumference.*

**420 boys were excluded because they met the inclusion criteria.

MPA=moderate physical activity.

VPA=vigorous physical activity.

MVPA=moderate and vigorous physical activity.

PA=physical activity.

SB=sedentary behavior.

Significant associations are in bold.

BOTTOM: The trend in girls of the HELENA study ():

Independents Variables	Null model		Unadjusted		Adjusted*	
	HELENA	BRACAH	HELENA	BRACAH	HELENA	BRACAH
	ß (95% CI)	ß (95% CI)	ß (95% CI)	ß (95% CI)	ß (95% CI)	ß (95% CI)
Fixed Effects Constant	117.6 (115.7–119.6)	107.0 (101.1–108.0)			90.4 (83.4–97.5)	77.2 (66.9–87.5)

Physical activity by questionnaire			p=0.844	p=0.054	p=0.443	p=0.099
≥60 min/d			Ref	Ref	Ref	Ref
<60 min/d			−0.15 (−1.62–1.33)	2.11 (−0.04–4.27)	−0.64 (−2.38–1.11)	1.63 (−0.50–3.77)
Sedentary behavior by questionnaire			p=0.130	p=0.402	p=0.963	p=0.501
<2 h/d			Ref	Ref	Ref	Ref
2–4 h/d			0.43 (−0.83–1.70)	4.05 (−2.30–10.39)	−0.20(−1.63–1.24)	3.47 (−2.65–9.61)
>4 h/d			1.53 (0.04–3.03)	3.95 (−1.83–9.73)	0.05 (−2.39–1.82)	2.98 (−2.59–8.55)
Physical activity by accelerometers **			p=0.635		p=0.818	
≥60 min/d			Ref		Ref	
<60 min/d			0.40 (−2.08–1.27)		0.20 (−1.99–1.53)	
MPA by accelerometers (min/d)			p=0.921		p=0.706	
		0.02 (−0.01–0.06)		−0.01 (−0.07–0.05)		
VPA by accelerometers (min/d)			p=0.16		p=0.176	
		−0.05 (−0.12–0.02)		−0.05 (−0.13–0.02)		
MPA by accelerometers (min/d)			p=0.484		p=0.364	

		−0.01 (−0.05–0.02)		−0.02 (−0.06–0.02)		
SB by accelerometers (min/d)			p=0.328		p=0.331	
		0.01 (−0.01–0.01)		0.01 (−0.01–0.01)		
Cluster PA and SB by accelerometers *			p=0.876		p=0.979	
<60 min/d+3rd tertile			Ref		Ref	
<60 min/d+2nd tertile			0.35 (−1.68–3.38)		0.14 (−1.86–2.13)	
<60 min/d+1st tertile			0.48 (1.81–2.76)		−0.75 (−3.00–1.49)	
≥60 min/d+3rd tertile			**3.59 (0.44–6.74)**		**3.61 (0.39–6.82)**	
≥60 min/d+2nd tertile			−0.69 (−3.43–2.05)		−1.07 (−3.83–1.68)	
≥60 min/d+1st tertile			−0.01 (−2.74–2.72)		0.22 (−2.53–2.98)	
Intraclass correlation coeficient	0.007	0.06			0.10	0.05
Standard deviation context	3.42	2.87			4.03	2.57
Standard deviation individual	12.75	11.4			11.88	10.92
Akaike Information Criterion	13.442.2	4123.9			10,978.2	4,115.1

Beta coefficient and their respective confidence intervals 95% (95% CI).

*This analysis was adjusted for potential confounders: *age, socioeconomic status, parental education, regular tobacco smoking, body mass index and waist circumference.*

**622 girls were excluded because they did not meet the inclusion criteria.

MPA=moderate physical activity.

VPA=vigorous physical activity.

MVPA=moderate and vigorous physical activity.

PA=physical activity.

SB=sedentary behavior.

Significant associations are in bold.

A cross-sectional study by Durrani and Fatima (2015) assessed the impact of physical activity on 701 school children in Aligarh city, India. The classification was on the following basis: the blood pressure between 60 to 75 mmHg (DBP) and 76 to 100 mmHg (SBP): equal to or below 25th percentile, was considered the low risk pressure group (LBP) while 85 to 100 mmHg (DBP) and 120 to 140 mmHg (SBP), equal to or greater than 95th percentile was considered the high risk or elevated pressure group (EBP).

Table 23 below is demonstrative of the ***TOP***: physical activity across the low/elevated blood pressure groups (Copyright © 2015 Anisa M. Durrani and Waseem Fatima. *This is an open access article distributed under the Creative Commons Attribution License, which permits unrestricted use, distribution, and reproduction in any medium, provided the original work is properly cited*):

TABLE 2: Mean and standard deviation of one-day physical activity (hrs/day) and energy expenditure in both boys and girls of LBP and EBP groups.

Physical activity (hrs)	LBP ($n = 50$)	EBP ($n = 50$)	Z
sleep	8.21 (1.22)	8.28 (1.15)	0.60
Very light	1.58 (1.1)	1.59 (1.27)	0.50
Light	22.46 (5.3)	21.72 (4.9)	0.080
Moderate	4.2 (3.3)	2.37 (2.57)	1.9**
Strenuous	1.15 (2.3)	1.63 (2.56)	0.60
Total activity score	37.67 (4.4)	35.41 (6.05)	1.4*

Figures in parenthesis indicate standard deviation.
*$P < 0.05$.
**$P < 0.001$.

The children of the LBP group displayed more moderate physical activity vs. the EBP group. While the total activity of the former was 37.67 ± 4.4 that of the latter was 35.41 ± 6.05. Further, the diastolic blood pressure of boys of the EBP group demonstrated a positive association with moderate activity and energy expenditure.

***BOTTOM*:** Correlation analyses of physical activity and BP (Copyright © 2015 Anisa M. Durrani and Waseem Fatima. *This is an open access article distributed under the Creative Commons Attribution License, which permits unrestricted use, distribution, and reproduction in any medium, provided the original work is properly cited*):

TABLE 3: Correlation coefficient of blood pressure of LBP and EBP group in boys and girls by physical activity and energy expenditure.

Physical activity	LBP ($n = 50$)				EBP ($n = 50$)			
	DBP		SBP		DBP		SBP	
	Girls (25)	Boys (25)	Girls (25)	Boys (25)	Girls (25)	Boys (25)	Girls (25)	Boys (25)
Sleep	0.00	−0.17	0.21	−0.01	−0.01	−0.11	−0.30	−0.36
Very light	0.10	−0.07	−0.07	−0.39	−0.20	−0.14	−0.10	0.25
Light	−0.10	−0.02	−0.02	−0.22	−0.18	0.21	0.17	0.34
Moderate	−0.17	−0.16	0.04	−0.01	−0.01	0.42*	−0.08	0.27
Strenuous	0.48*	0.19	0.13	0.24	0.17	−0.09	−0.06	−0.30
Total activity score	−0.10	−0.05	0.09	0.30	−0.14	0.23	0.09	0.16

Figures in parenthesis indicate number of boys and girls.
* $P < 0.05$.

2.9 FINAL POINTS

A snapshot of recommendations to be adopted for the diet and lifestyle for children with high blood pressure is presented below in table 24 (Spagnolo et al, 2013: *Open Access: This article is published under license to BioMed Central Ltd. This is an Open Access article is distributed under the terms of the Creative Commons Attribution License (https://creativecommons.org/licenses/by/2.0),* which permits unrestricted use, distribution, and reproduction in any medium, provided the *original work is properly cited*):

Goals	
•	**BMI in the normal weight range**: maintain BMI to prevent overweight
•	**BMI in the overweight range**: weight maintenance in younger children or gradual weight loss in older children and adolescents to return to the normal weight range
•	**BMI in the obesity range**: gradual weight loss (1–2 kg/month) to achieve normal weight
General recommen-dations	
•	Moderate to vigorous physical aerobic activity for 40 min, 3–5 days/week and avoid more than 2 h of daily sedentary activities (besides school hours)

•	Participation in (competitive) sports activities limited only in the presence of uncontrolled stage 2 hypertension
•	Avoid severe dietary restrictions, reduce portion size, stimulate the habit of having breakfast
•	Limit salt intake
•	Avoid intake of excess sugar, excess soft drinks, saturated and trans fat, animal protein
•	Drink water
•	Stimulate the intake of healthy food (fruit, vegetables, legumes, whole grain products, fish)
•	Implement the behavioral changes (physical activity and diet) suitable for individual and family characteristics
•	Establish realistic goals
•	Involve the family, caregivers, teachers and other educators in the process of dietary and life-style changes
•	Provide educational support and materials
•	Develop reward systems (non-food) to achieve healthy behavior

Further, given the importance of optimal physical activity, table 25 below presents recreational activities for children or adolescents (***under medical guidance***) (Strambi et al, 2016: *Open Access: This article is distributed under the terms of the Creative Commons Attribution 4.0 International License (http://creativecommons.org/licenses/by/4.0/), which permits unrestricted use, distribution, and reproduction in any medium, provided you give appropriate credit to the original author(s) and the source, provide a link to the Creative Commons license, and indicate if changes were made. The Creative Commons Public Domain Dedication waiver (http://creativecommons.org/publicdomain/zero/1.0/) applies to the data made available in this article, unless otherwise stated*):

Mountaineering, Climbing
Athletics: speed, jumps, leaps, heptathlon, decathlon
Bobsledding, Luge
Weight lifting, Bodybuidling
Speed cycling, Mountain bike downhill, BMX
Artistic gymnastics
Synchronized swimming

Motorcycling (Motocross, Enduro, Trial)
Water skiing
Skiing: slalom, giant slalom, super G, downhill, alpine, snowboard, carving, grass skiing.
Surfing
Tug-of-war

Please note that the content given represents information and is not to be used for diagnosis and treatment without the advice of a qualified healthcare provider.

"The adolescent must never be treated as a child, for that is a stage of life that he has surpassed. It is better to treat an adolescent as if he had greater value than he actually shows than as if he had less and let him feel that his merits and self-respect are disregarded"

-Maria Montessori

2.10 REFERENCES:

1. Spagnolo, A., Giussani, M., Ambruzzi, A.M. *et al.* Focus on prevention, diagnosis and treatment of hypertension in children and adolescents. *Ital J Pediatr* **39,** 20 (2013). https://doi.org/10.1186/1824-7288-39-20
2. Hardy ST, Sakhuja S, Jaeger BC, et al. Trends in Blood Pressure and Hypertension Among US Children and Adolescents, 1999-2018. *JAMANetw Open.* 2021;4(4):e213917. doi:10.1001/jamanetworkopen.2021.3917
3. Daniel RA, Haldar P, Prasad M, Kant S, Krishnan A, Gupta SK, et al. (2020) Prevalence of hypertension among adolescents (10-19 years) in India: A systematic review and meta-analysis of cross-sectional studies. PLoS ONE 15(10): e0239929. https://doi.org/10.1371/journal.pone.0239929
4. Ejike, C.E., Ugwu, C.E. & Ezeanyika, L.U. Variations in the prevalence of point (pre)hypertension in a Nigerian school-going adolescent population living in a semi-urban and an urban area. *BMC Pediatr* **10,** 13 (2010). https://doi.org/10.1186/1471-2431-10-13.
5. Procolo Di Bonito and Anna Di Sessa. New Diagnostic Criteria for Hypertension in Children and Adolescents: Lights and Shadows. Children 2020, 7, 196; doi:10.3390/children7110196.
6. Ashraf, M., Irshad, M. & Parry, N.A. Pediatric hypertension: an updated review. *ClinHypertens* **26,** 22 (2020).https://doi.org/10.1186/s40885-020-00156-w
7. Sayeemuddin M, Sharma D, Pandita A, Sultana T and Shastri S (2015) Blood pressure profile in school children (6–16 years) of Southern India: a prospective observational study. *Front. Pediatr.* **3**:24. doi:10.3389/fped.2015.00024
8. Souza, CamilaBrandão de, Dourado, Cláudia de Souza, Mill, José Geraldo, Salaroli, LucianeBresciane, & Molina, Maria del Carmen Bisi. (2017). Prevalence of Hypertension in Children from Public Schools. *International Journal of Cardiovascular Sciences*, *30*(1), 42-51. https://doi.org/10.5935/2359-4802.20170023
9. Lu, X., Shi, P., Luo, CY. *et al.* Prevalence of hypertension in overweight and obese children from a large school-based population in Shanghai, China. *BMC Public Health* **13,** 24 (2013).https://doi.org/10.1186/1471-2458-13-24

10. Xiaohua Liang, Lun Xiao, YetaoLuo, JiapeiXu, "Prevalence and Risk Factors of Childhood Hypertension in Urban-Rural Areas of China: A Cross-Sectional Study", International Journal of Hypertension, vol. 2020, Article ID 2374231, 18 pages, 2020. https://doi.org/10.1155/2020/2374231
11. Orlando A, Cazzaniga E, Giussani M, Palestini P and Genovesi S (2018) Hypertension in Children: Role of Obesity, Simple Carbohydrates, and Uric Acid. *Front. Public Health* 6:129. doi: 10.3389/fpubh.2018.00129
12. Farhangi, M.A., Nikniaz, L. &Khodarahmi, M. Sugar-sweetened beverages increases the risk of hypertension among children and adolescence: a systematic review and dose–response meta-analysis. *J Transl Med* **18,** 344 (2020).https://doi.org/10.1186/s12967-020-02511-9
13. Kit BK, Fakhouri TH, Park S, Nielsen SJ, Ogden CL. Trends in sugar-sweetened beverage consumption among youth and adults in the United States: 1999-2010. Am J ClinNutr. 2013;98(1):180–8.
14. Yu H, Zou L, He Y, Luo L, Dong W, Zhang Y, et al. (2019) Associations between neonatal serum bilirubin and childhood hypertension. PLoS ONE 14(7): e0219942. https://doi.org/10.1371/journal.pone.0219942
15. Kuciene, R., Dulskiene, V. Associations of short sleep duration with prehypertension and hypertension among Lithuanian children and adolescents: a cross-sectional study. *BMC Public Health* **14,** 255 (2014). https://doi.org/10.1186/1471-2458-14-255
16. Gradisar M, Gardner G, Dohnt H. Recent worldwide sleep patterns and problems during adolescence: a review and meta-analysis of age, region, and sleep. Sleep Med. 2011 Feb;12(2):110-8. doi: 10.1016/j.sleep.2010.11.008. Epub 2011 Jan 22. PMID: 21257344.
17. Bal, C., Öztürk, A., Çiçek, B., Özdemir, A., Zararsız, G., Ünalan, D., Ertürk Zararsız, G., Korkmaz, S., Göksülük, D., Eldem, V., İsmailoğulları, S., Erdem, E., Mazıcıoğlu, M. M., & Kurtoğlu, S. (2018). The Relationship Between Blood Pressure and Sleep Duration in Turkish Children: A Cross-Sectional Study. *Journal of clinical research in pediatric endocrinology, 10*(1), 51–58. https://doi.org/10.4274/jcrpe.4557
18. LiliHuang, JiajunLyu, ZichongLong, YuanqingXia, YitingChen, Xiuxia Ye, Shenghui Li, "Gender-Specific Longitudinal Association of Sleep Duration with Blood Pressure among Children: Evidence from CHNS

2004–2015", International Journal of Hypertension, vol. 2020, Article ID 5475297, 9 pages, 2020. https://doi.org/10.1155/2020/5475297

19. Ziegelasch, N., Vogel, M., Siekmeyer, W. *et al.* Seasonal variation of blood pressure in children. *Pediatr Nephrol* (2020). https://doi.org/10.1007/s00467-020-04823-w
20. de Moraes ACF, Carvalho HB, Rey-López JP, Gracia-Marco L, Beghin L, Kafatos A, et al. (2013) Independent and Combined Effects of Physical Activity and Sedentary Behavior on Blood Pressure in Adolescents: Gender Differences in Two Cross-Sectional Studies. PLoS ONE 8(5): e62006. https://doi.org/10.1371/journal.pone.0062006
21. Anisa M. Durrani, Waseem Fatima, "Effect of Physical Activity on Blood Pressure Distribution among School Children", Advances in Public Health, vol. 2015, Article ID 379314, 4 pages, 2015. https://doi.org/10.1155/2015/379314
22. Strambi, M., Giussani, M., Ambruzzi, M.A. *et al.* Novelty in hypertension in children and adolescents: focus on hypertension during the first year of life, use and interpretation of ambulatory blood pressure monitoring, role of physical activity in prevention and treatment, simple carbohydrates and uric acid as risk factors. *Ital J Pediatr* **42,** 69 (2016). https://doi.org/10.1186/s13052-016-0277-0

Chapter 3

Controlling Hypertension and Research Insights: A Glimpse

Shivsanjeevi Sripathi

Author's note: This chapter is not meant to be as medical advice or to be considered as a treatment option. Please consult a healthcare giver or center for relevant advice.

"Tell me what you eat, and I will tell you who you are".

–Jean Anthelme Brillat-Savarin

CONTENTS

3.1 Diet 112

3.2 miRNAs of the Gut-Liver Axis 133

3.3 References 142

3.1 DIET

The 1990s saw the emergence of the Dietary Approaches to Stop Hypertension (DASH) diet. Early work by the National Institute of Health (NIH)-funded projects resulted in a dip in the systolic blood pressure by 6 to 11 mm Hg that led to the advocacy of the diet to control hypertension. The focus of the diet is to limit the eating of processed foods with more fresh foods, lowering sodium to 1500 mg/day in food, eating fruits and vegetables, adjustment of micronutrients and lean meat inclusion for meat eaters. The guide is listed below:

- Vegetables: around five servings per day
- Fruits: around five meals per day
- Carbohydrates: close to seven servings per day
- Low-fat dairy products: approximately two servings per day
- Lean meat products: around two or lesser servings per day
- Nuts and seeds: twice or thrice a week

In this diet, ***proteins*** are inclusive of plant-based proteins like *legumes, nuts, soy and seeds.* Animal sources include *lean meats, low-fat dairy, fish and eggs*. The recommended ***carbohydrates*** are *whole grains, green leafy vegetables, beans, legumes and low glycemic index fruits*. The recommended ***fats*** are *nuts, flax seeds, avocados, olive oil and omega-3 fatty acids*.

Micronutrients have also been recommended such as *potassium* (present in bananas, spinach and oranges), *magnesium* (seeds, nuts, green leafy vegetables and whole grains) and *calcium* (green leafy vegetables and whole milk).

(Challa HJ, Ameer MA, Uppaluri KR: Bookshelf ID: NBK482514: PMID: 29494120;

In order to understand the impact of this diet, a 2020-published study in the *Nutrition Journal* by Soltani and team (2020) described as a "*systematic review and dose-response meta-analysis of prospective observational studies, for the first time*" explored how all-cause mortality was impacted by adherence to the DASH diet. The study (registration ID: CRD42018086500 in the international prospective register of systematic reviews (PROSPERO) database) entailed the analyses of the following studies that were narrowed down by the authors displayed in table 1 by the study (Soltani et al, 2020: *Open Access: This article is licensed under a Creative Commons Attribution 4.0 International License, which permits use, sharing, adaptation, distribution and reproduction in any medium or format, as long as you give appropriate credit to the original author(s) and the source, provide a link to the Creative Commons licence, and indicate if changes were made. The images or other third party material in this article are included in the article's Creative Commons licence, unless indicated otherwise in a credit line to the material. If material is not included in the article's Creative Commons licence and your intended use is not permitted by statutory regulation or exceeds the permitted use, you will need to obtain permission directly from the copyright holder. To view a copy of this licence, visit http://creativecommons.org/licenses/by/4.0/. The Creative Commons Public Domain Dedication waiver (http://creativecommons.org/publicdomain/zero/1.0/) applies to the data made available in this article, unless otherwise stated in a credit line to the data*):

Key: [1]Food frequency questionnaire

Author name, publication year	**Study name, country**	**Follow-up duration (year)**	**Mortality endpoints (Num)**	**Diet assessment method, item, assessment time**	**DASH dietary pattern score range/ component (score range)**	**Covariates**
Folsom 2007 [37]	Iowa Women s Health Study, USA	16	CVD (1121) Stroke (236)	Validated FFQ[1], 127, baseline	11 components: Total grains, whole grains, fruits, vegetables, total dairy, total meats, nuts, seeds and dry beans, sweets, sodium, total fats, saturated fatty acids. (8–40)	Age, energy intake, education, BMI, waist/hip, smoking status and pack-years, estrogen use, alcohol intake, physical activity, and multivitamin use

George 2014 [22]	Women's Health Initiative Study, USA	12.9	CVD (1483) Cancer (2384)	Validated FFQ, 122, baseline	8 components: fruit, vegetables, nuts and legumes, whole grains, low-fat dairy products, sodium, red and processed meats and sweetened beverages (8–40)	Age, race/ethnicity, educational level, marital status, smoking, physical activity, daily energy intake, postmenopausal hormone therapy, diabetes, BMI, and alcohol
Yu 2014 [34]	Shanghai Men's Health Study, China	6.5	All-cause (4348)	Validated FFQ, 81, baseline	7 components: vegetables; fruits; dairy; meat, poultry, fish, and eggs; nuts, seeds, and legumes; fats and oils; and sodium. (0–70)	Age, Educational Attainment, Income, Cigarette smoking, Alcohol consumption, Physical activity, multivitamin supplement; BMI, Waist-to-hip ratio; History of CVD, Diabetes, or Hypertension; and Total energy intake.
			CVD (1344)			
			Cancer (1836)			
	Shanghai Women's Health Study, China	12	All-cause (2954)	FFQ, 77, baseline		
			CVD (964)			
			Cancer (1290)			
Reedy 2014 [38]	NIH-AARP Diet and Health Study, USA	15	All-cause (54871)	Validated FFQ, 124, baseline	8 components: fruit, vegetables, nuts and legumes, whole grains, low-fat dairy products, sodium, red and processed meats, and sweetened beverages. (8–40)	Age, ethnicity, education, BMI, smoking, vigorous physical activity, energy intake, marital status, Diabetes, and alcohol
			CVD (15497)			
			Cancer (18646)			

Cuenca-García 2014 [36]	Aerobics Center Longitudinal Study, USA	11.6	All-cause (358) CVD (102) Cancer (134)	3-day diet record, baseline	8 components: fruits and vegetables, fish, whole grains, nuts/seeds and legumes, sodium, added sugar, processed meat, and saturated fat. (0–8)	Age, sex, energy intake, examination year, smoking, alcohol, abnormal resting or exercise electrocardiogram, parental history of premature CVD, physical activity, and cardiorespiratory fitness
Boggs 2015 [39]	Black Women s Health Study, USA	16	All-cause (1678) CVD (428) Cancer (742)	Validated FFQ, 98, baseline and every 2 year	8 components: fruits (including fruit juice), vegetables, nuts and legumes, whole grains, and low-fat dairy, sodium, red and processed meats, and sugar-sweetened beverages. (8–40)	Age, total energy intake, education, marital status, vigorous exercise, television watching, smoking, BMI, and alcohol
Harmon 2015 [23]	Multiethnic Cohort Study, USA	13–18	All-cause (18263) CVD (6408) Cancer (5853)	Validated FFQ, 182, baseline	8 components: fruit, vegetables, nuts and legumes, whole grains, low-fat dairy products, sodium, red and processed meats, and sweetened beverages. (8–40)	Age, energy intake, history of diabetes, ethnicity, moderate-to-vigorous physical activity, smoking, education, marital status, hormone-replacement therapy, BMI, and alcohol

Lassale 2016 [40]	Pan-EPIC, Europe	12.8	All-cause (15200) CVD (3761) Cancer (7475)	Validated FFQ, baseline	8 components: fruit, vegetables, nuts and legumes, whole grains, low-fat dairy products, sodium, red and processed meats, and sweetened beverages. (8–40)	Age, sex, BMI, smoking status, physical activity level and educational level
Park 2016 [41]	Third National Health and Nutrition Examination Survey	18.6	All-cause (296) CVD (67) Cancer (79)	A single 24-h dietary recall, baseline	9 components: fruits, vegetables, nuts and legumes, low-fat dairy products, whole grains, fat (total/saturated), sodium, sweets, and red meats. (0–9)	Age, sex, race/ethnicity, educational attainment, income, smoking status, alcohol consumption, physical activity, and total energy intake
Biesbroek 2017 [42]	EPIC-NL, Netherland	19.2	All-cause (3845)	Validated FFQ, 178, baseline	8 components: fruit, vegetables, nuts and legumes, whole grains, low-fat dairy products, sodium, red and processed meats, and sweetened beverages. (8–40)	Age, BMI, educational level, smoking status, total energy intake, physical activity and alcohol consumption
Zaslavsky 2017 [33]	Women s Health Initiative Study, USA	12.4	All-cause (3259)	Validate FFQ, 122, baseline and 3 year after baseline	8 components: fruit, vegetables, nuts and legumes, whole grains, low-fat dairy products, sodium, red and processed meats, and sweetened beverages. (8–40)	Age, energy and protein intake, race/ethnicity, education, income, smoking, number of frailty criteria, physical activity, and BMI

Jones 2018 [43]	EPI C-Norfo lk cohor t study, Britania	12.4	All-cause (6567)	Validated FFQ, 130, baseline	8 components: fruit, vegetables, nuts and legumes, whole grains, low-fat dairy, red and processed meats, salt, and non-milk extrinsic sugars. (8–40)	Age, sex and dietary energy, smoking status, alcohol intake, physical activity, BMI, diabetes, SES, marital status, use of anti-hypertensive medication, use of lipid-lowering medication and history of CVD
			CVD (1647)			
Aigner 2018 [44]	Multi-ethnic Cohort Study, USA	17.6	Stroke (3548)	Validated FFQ, 180, baseline	8 components: fruit, vegetables, nuts and legumes, whole grains, low-fat dairy products, sodium, red and processed meats, and sweetened beverages. (8–40)	Age, energy intake, history of diabetes, ethnicity, moderate-to-vigorous physical activity, smoking, education, marital status, hormone-replacement therapy, BMI, and alcohol

The researchers reported an inverse-dose-response association between death and the DASH diet with clinical implications given that the authors stated that an increase in adherence by five points to the diet resulted in a dip in mortality scores vs. non-adherence to the diet as displayed below in figure 1 (Soltani et al, 2020: *Open Access: This article is licensed under a Creative Commons Attribution 4.0 International License, which permits use, sharing, adaptation, distribution and reproduction in any medium or format, as long as you give appropriate credit to the original author(s) and the source, provide a link to the Creative Commons licence, and indicate if changes were made. The images or other third party material in this article are included in the article's Creative Commons licence, unless indicated otherwise in a credit line to the material. If material is not included in the article's Creative Commons licence and your intended use is not permitted by statutory regulation or exceeds the permitted use, you will need to obtain permission directly from the copyright holder. To view a copy of this licence, visit http://creativecommons.org/licenses/by/4.0/. The Creative*

Commons Public Domain Dedication waiver (http://creativecommons.org/publicdomain/zero/1.0/) applies to the data made available in this article, unless otherwise stated in a credit line to the data):

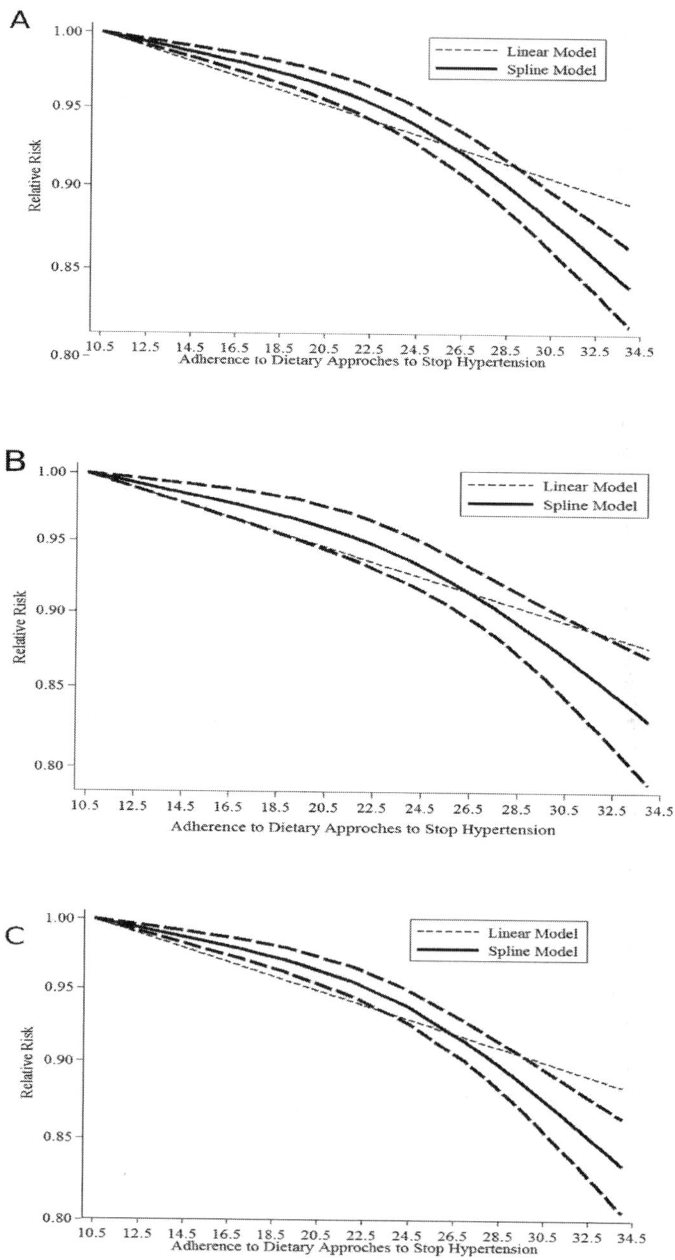

As mentioned above, the association of hypertension and micronutrients has been described by Chiu et al (2021) in *Nutrients*. Various minerals in their ionized form work in tandem with other macronutrients to impact osmotic pressure, acid-base balance and muscle movements. This is depicted as below in figure 2 (Chiu, H. F., Venkatakrishnan, K., Golovinskaia, O., & Wang, C. K., 2021: Copyright © 2021 by the authors. Licensee MDPI, Basel, Switzerland. *This article is an open access article distributed under the terms and conditions of the Creative Commons Attribution (CC BY) license: http://creativecommons.org/licenses/by/4.0/*):

Key: SNS—sympathetic nervous system; RAAS—renin-angiotensin-aldosterone system; Na+—sodium; K+—potassium; Ca2+—calcium; Mg2+—magnesium; Zn2+—zinc; Cu2+—copper; Se2—selenium

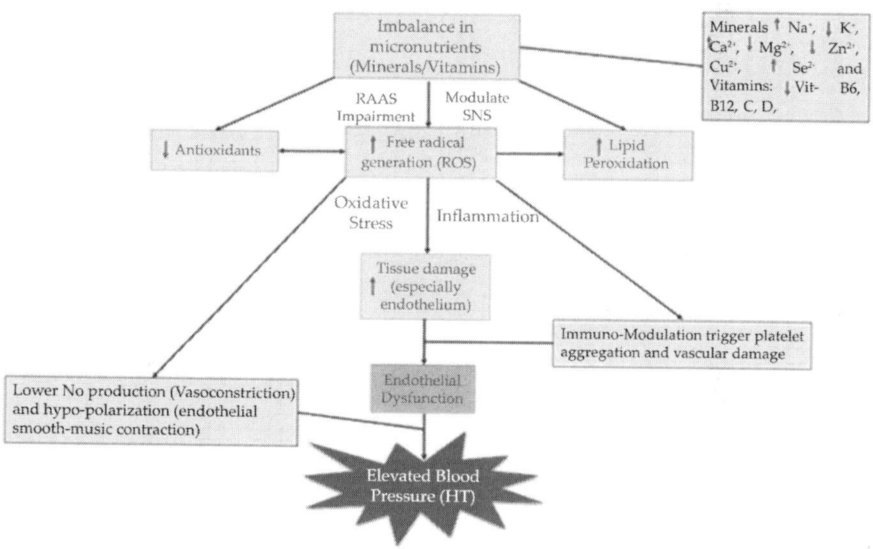

The impact of different electrolytes on hypertension was compiled in a summary of meta-analyses by Iqbal, S., Klammer, N., & Ekmekcioglu, C (2019) in *Nutrients*. For **sodium**, table 2 below depicts the summary (Iqbal, S., Klammer, N., & Ekmekcioglu, C., 2019: Copyright © 2019 by the authors: Licensee MDPI, Basel, Switzerland. *This article is an open access article distributed under the terms and conditions of the Creative Commons Attribution (CC BY) license http://creativecommons.org/licenses/by/4.0/*):

Key: SBP = Systolic blood pressure; DBP = Diastolic blood pressure

Author/ Year	No. of Trials	Study Characteristics	No. of Participants	Patient Characteristics	Duration of Trials	Sodium/Salt Intake or Reduction	Blood Pressure Lowering in mmHg (95% CI)	Further Remarks/ Summary
Aburto et al., 2013a [24]	36	Randomized controlled trials	6736	2273 (hypertensive)	Most studies (n = 31) <3 months	Different reductions in sodium intake Relative sodium reduction in the intervention group: ≥1/3 of control	SBP: −3.4 (−4.3 to −2.5) DBP: −1.5 (−2.1 to −1.0)	Reduced sodium intake decreases blood pressure in people both with and without hypertension. The reduction in blood pressure was greater in those with hypertension.
Adler et al., 2014 [25]	6 (SBP) 5 (DBP)	Randomized controlled trials	3362 (SBP) 2754 (DBP)	SBP (end of trial): (normotensive) 2079 (hypertensive) 1283 DBP (end of trial): (normotensive) 2079 (hypertensive) 675	7–36 months	Sodium intake: 70 to >100 mmol/day	Normotensive: SBP: −1.2 (−2.3 to 0.02) DBP: −0.8 (−1.4 to −0.2) Hypertensive: SBP: −4.1 (−5.8 to −2.4) DBP: −3.7 (−8.4 to 0.9)	Normotensive persons: small blood pressure reduction. Hypertensive patients: greater reduction in SBP, no difference in DBP.
Graudal et al., 2015 [26]	15	Randomized controlled trials	12–114	"time to maximal efficacy" analysis 7 studies with hypertensive patients 7 studies with normotensive persons 1 study hypertensive + normotensive	1 to 6 weeks	Sodium reduction range: 55–118 mmol/day	No significant differences in SBP or DBP after initiation of salt reduction between week 1 and subsequent weeks.	Time dependent effects of salt reduction on blood pressure. The effect of salt reduction on blood pressure appears to reach maximal efficacy at 1 week and remain stable over subsequent time intervals.

Graudal and Jürgens 2015 [27]	92	Randomized controlled trials	661 Asians 561 Blacks 3782 Whites	9 Asian/9 Black/74 White population	7–365 days	Sodium reduction: 63–103 mmol	SBP: −3.2 (−4.0 to −2.5) (in Whites) −4.7 (−7.1 to −2.3) (in Blacks) −3.8 (−6.4 to −1.3) (in Asians) DBP:−1.5 (−2.1 to −1.0) (in Whites) −3.0 (−4.0 to −2.0) (in Blacks) −2.0 (−3.0 to −0.9) (in Asians)	SBP: no differences in ethnic groups. DBP: small differences between black and white people.
Gay et al., 2016 [28]	24	Randomized controlled trials	23,858	11 to 2570 participants (median: 129) participants >19 years old	Trial durations ranged from 6 to 48 months of follow–up (median: 12 months)	Dietary interventions (including low sodium diets)	Overall pooled net effect of diets: SBP: −3.1 (−3.9 to −2.3) DBP: −1.8 (−2.2 to −1.4)	This meta-analysis shows that dietary interventions (including low sodium diets) provide clinically significant net blood pressure reductions, and that some dietary patterns may be more effective than others.

Graudal et al., 2017 [29]	177	Random-ized controlled trials	12,210	White people with hypertension. (84 studies; 5925 participants in SBP; 85 studies; 6001 participants in DBP) Black people with hypertension. (8 studies; 619 participants in SBP and DBP) Asian people with hypertension. (8 studies; 501 participants in SBP and DBP) White people with normotension. (89 studies, 8569 participants in SBP; 90 studies, 8833 participants in DBP) Black people with normotension. (7 studies, 506 participants in SBP and DBP) Asian people with normotension. (3 studies, 393 participants in SBP and DBP)	4–1100 days	Mean sodium reduction: 135 mmol/day range; <100 to ≥250 mmol/day	Hypertensive (White): SBP: −5.5 (−6.5 to −4.6) DBP: −2.9 (−3.4 to −2.3) Hypertensive (Black): SBP: −6.6 (−9.0 to −4.2) DBP: −2.9 (−4.5 to −1.30) Hypertensive (Asian): SBP: −7.8 (−11.4 to −4.1) DBP: −2.7 (−4.2 to −1.2) Normotensive (White): SBP: −1.1 (−1.6 to −0.6) DBP: 0.03 (−0.4 to 0.4) Normotensive (Black) SBP: −4.0 (−7.4 to −0.7) DBP: −2.0 (−4.4 to 0.4) Normotensive (Asian): SBP: −0.7 (−3.9 to 2.4) DBP: −1.6 (−3.4 to 0.1)	High-quality evidence for White people; moderate-quality evidence for Black/Asian people

He et al., 2013 [30]	34	Randomized trials	3230	990 (of 22 trials) hypertensive 2240 (of 12 trials) normotensive	Median duration: 5 weeks in hypertensive people, 4 weeks in normotensive people	Salt reduction: 75 mmol/day (4.4 g/day). Reduction of urinary sodium: 40–120 mmol/day (2.3–7.0 g/day).	Total SBP: −4.2 (−5.2 to −3.2) DBP: −2.1 (−2.7 to −1.5) Hypertensive: SBP: −5.4 (−6.6 to −4.2) DBP: −2.8 (−3.5 to −2.1) Normotensive: SBP: −2.4 (−3.6 to −1.3) DBP: −1.0 (−1.9 to −0.2)	Reduction in SBP was significant in both black and white people and in women and men. Significant effects on blood pressure were seen in hypertensives and normotensives. Dose-response relation: the greater the reduction in salt intake, the greater the fall in blood pressure.
He and MacGregor 2011 [31]	6	Outcome trials	6250	3 trials in normotensive participants 3 trials in hypertensive patients	6–36 months	Salt reduction: 2–2.3 g/day	Normotensive: SBP: −1.1 (−0.1 to 2.3) DBP: −0.8 (0.2 to 1.4) Hypertensive: SBP: −4.1 (2.4 to 5.8) DBP: −3.7 (−0.9 to 8.4)	Significant reduction in cardiovascular events
Kelly et al., 2016 [32]	5	Randomized and non-randomized controlled trials	1214	Normotensive participants (≥18 years) with SBP ≤140 mmHg	4 weeks to 48 months	Salt reduction: −75 mmol/day (range; −37 to −136 mmol).	SBP: −0.7 (−2.6 to 1.2) DBP: −0.6 (−1.3 to 0.1)	No significant change in SBP or DBP following reduction of dietary sodium over the period of 4 weeks to 36 months
Peng et al., 2014 [33]	5	Randomized controlled trials	1974	Hypertensive and normotensive participants	6 months to 2 years	Different salt substitutes vs. common salt (NaCl).	SBP: −4.9 (−7.3 to −2.5) DBP: −1.5 (−2.7 to −0.3)	Salt substitutes significantly reduced both SBP and DBP

Taylor et al., 2011 [34]	7	Randomized controlled trials	3 trials normotensive (3518), 2 trials hypertensive (758), 1 trial mixed pop. (1981), 1 trial with heart failure (234)	Adults ≥18 years, irrespective of gender/ethnicity. Studies of children/pregnant women were excluded.	Trials follow-up ranged 6 to 71 months	Salt reduction; <70–100 mmol/day. Urinary 24-h sodium excretion: Normotensive (mean diff.): 34.2 mmol/24 h (18.8–49.6), Hypertensive (mean diff.): 39.1 mmol/24 h (31.1–47.1)	Normotensives (mean difference) SBP:−1.1 (−2.3 to 0.1) DBP:−0.8 (−1.4 to −0.2) Hypertensives SBP: −4.1 (−5.8 to −2.4) DBP: −3.7 (−8.4 to 0.9)	Significant reduction of SBP in hypertensive patients
Wang et al., 2015 [35]	6	Interventional studies	3153	Chinese adults aged ≥35 years	At most 1 week	Salt level reduced in hypertensive patients: 9.6 g/day (163.0 mmol/day sodium).	Normotensive + hypertensive: SBP: −6.3 (−7.2 to −5.4) DBP: −3.2 (−3.7 to −2.7) Hypertensive: SBP: −8.9 (−14.1 to −3.7) DBP: −5.9 (−9.7 to −2.1)	Salt restriction lowers mean BP in Chinese adults, with the strongest effect among hypertensive participants.
Observational Studies								

Talukder et al., 2017 [36]	10	Obser-vational studies	8093	7 studies (12 datasets) with 3747 participants with low/high water sodium exposure groups		4–405 mg/L water sodium level	Standardized mean difference: SBP: 0.1 (−0.2 to 0.3) DBP: 0.2 (0.1 to 0.4)	An (inconclusive) association between water sodium and human blood pres-sure is suggested, more consistently for DBP.
Subasinghe et al., 2016 [37]	18	Obser-vational studies	134,916	Participants in urban and rural areas in low-and-middle income countries (LMICs). Age: 24–65.		Daily salt intake range: 6.9 to 42.3 g/day	Effect size (ES) of hypertension ES 1.36 (1.24 to 1.48) ES 1.28 (1.13 to 1.45)	Excessive salt intake has a greater impact on the prevalence of hypertension in urban than rural regions.

Table 3 below summarizes the findings for the role of **calcium** on hypertension (Iqbal, S., Klammer, N., & Ekmekcioglu, C., 2019: Copyright © 2019 by the authors: Licensee MDPI, Basel, Switzerland. *This article is an open access article distributed under the terms and conditions of the Creative Commons Attribution (CC BY) license http://creativecommons.org/licenses/by/4.0/)*:

Author/ Year	**No. of Trials**	**Study Charac-teristic**	**No. of Partici-pants**	**Study Aims**	**Duration of Trials**	**Calcium Dosage (Diet or Supplement)**	**Blood Pressure Low-ering in mmHg or RR/ OR for Gestational Hypertension (95% CI)**	**Further Re-marks/Summary**
Imdad et al., 2011 [8]	6	Random-ized controlled trials	Calcium-group: 4919 Control group: 4942	Effect of cal-cium supple-mentation on gestational hypertensive disorders in studies from developing countries	Calcium supplements in all the in-cluded studies were before 20–32 weeks of gestation and continued till delivery.	0.5–2 g/day	RR: 0.55 (0.36 to 0.85)	Calcium supple-mentation during pregnancy was associated with a significant reduced risk of acquiring gesta-tional hyperten-sion.

Hofmeyr et al., 2014 [43]	12 trials	Randomized controlled trials	15,470 women	Assessing the effects of calcium supplementation during pregnancy on hypertensive disorders of pregnancy and related maternal and child outcomes	Calcium supplementation started at the latest from 34 weeks of pregnancy.	High-dose calcium supplementation (≥1 g/day)	RR: 0.65 (0.53 to 0.81)	Average risk of high blood pressure was reduced with calcium supplementation compared with placebo. There was also a reduction in hypertension with low-dose calcium supplementation (<1 g/day).
An et al., 2015 [44]	4	Randomized controlled trials	Gestational hypertension: 7252 Control group: 7272 Severe gestational hypertension: 6673 Control group: 6684	Assessing the effectiveness of calcium supplementation during pregnancy on reducing the risk of hypertensive disorders of pregnancy and related problems.	From ~11–24 weeks of pregnancy to delivery	Supplementation with calcium (at least >1 g/day)	Gestational hypertension: RR: 0.91 (0.84 to 0.99) Severe gestational hypertension: RR: 0.81 (0.60 to 1.09)	Calcium supplementation appears to reduce the risk of hypertension in pregnancy. No significant reduction in the risk of severe gestational hypertension.

Wu and Sun 2017 [46]	8	Randomized controlled trials	36,806	Evaluation the effect of calcium plus vitamin-D (CaD) supplements on the changes in BP from baseline to the longest follow-up time point in male and female participants.	8 weeks to 7 years	Intervention dose of calcium (≤1000 mg/day, 5 trials or >1000 mg/day, 3 trials)	Mean differences in SBP: 0.6 (−1 to 2.20) Mean differences in DBP: −0.2 (−0.9 to 0.5)	Calcium plus vitamin D supplementation slightly increased SBP, but the difference was not statistically significant. Calcium plus vitamin D supplementation did not significantly affected DBP reduction.
Cormick et al., 2015 [47]	16	Randomized controlled trials	SBP: 3048 (16 studies) DBP: 2947 (15 studies)	Assessing the efficacy and safety of calcium supplementation versus placebo or control for reducing blood pressure in normotensive people	Median follow up intervention period of 3.5 months	For most studies the intervention was 1000 mg to 2000 mg of elemental calcium per day	Mean difference: SBP: −1.4 (−2.2 to −0.7) DBP: −1 (−1.5 to −0.5)	The quality of evidence was high for doses of calcium of 1000 to 1500 mg/day and was moderate for lower or higher doses. Calcium intake slightly reduced both SBP and DBP in normotensive people.
Observational studies								

Schoenaker et al., 2014 [45]	16	Observational studies	Case-control studies: 757 pregnant women Cohort studies: 41,214 pregnant women, 908 gestational hypertension	Assessing the effect of dietary factors on hypertensive disorders during pregnancy (gestational hypertension and preeclampsia)		Highest group >1600 mg/day versus lowest group <1000 mg/day	Gestational hypertension (comparing highest to lowest): OR: 0.63 (0.41 to 0.97)	Results from case–control studies consistently showed lower reported calcium intake for pregnant women with hypertensive disorders (gestational hypertension and preeclampsia)

Table 4 below shows the studies assessing the level of **potassium** and hypertension (Iqbal, S., Klammer, N., & Ekmekcioglu, C., 2019: Copyright © 2019 by the authors: Licensee MDPI, Basel, Switzerland. *This article is an open access article distributed under the terms and conditions of the Creative Commons Attribution (CC BY) license http://creativecommons.org/licenses/by/4.0/)*:

Author/ Year	**No. of trials**	**Study Characteristic**	**No. of Participants**	**Patient Characteristics**	**Duration of Trials**	**Potassium Dosage (Supplements)**	**Blood Pressure Lowering in mmHg (95%CI)**	**Further Remarks/ Summary**
Aburto et al., 2013b [38]	21	Randomized controlled trials	1892/1857	Hypertensive 818(SBP)/828(DBP)	<2 to >4 months	<90 mmol/day to >155 mmol/day in the intervention group	SBP: −3.5 (−5.2 to −1.8) DBP: −2.0 (−3.1 to −0.9)	Effect seen in people with hypertension but not in those without hypertension. Intake above 120 mmol/day did not seem to have any additional benefit.

								Potassium may be more effective in reducing blood pressure at higher levels of sodium consumption.
Binia et al., 2015 [39]	15	Randomized controlled trials	917	400 hypertensives 329 normotensives 188 hypertensive or normotensive persons (mixed population)	4–24 weeks	<40–120 mmol/day	All: SBP: −4.7 (−7.0 to −2.4) DBP: −3.5 (−5.7 to −1.3) Hypertensive patients: SBP: −6.8 (−9.3 to −4.3) DBP: −4.7 (−7.5 to −1.8)	Potassium supplementation is associated with reduction of blood pressure in patients who are not on antihypertensive medication, and the effect is significant in hypertensive patients.
Filippini et al., 2017 [40]	33	Randomized controlled trials	1829	1163 (studies ≥4 weeks overall)	<4 to ≥12 weeks	25–250 mmol/day	SBP: −4.5 (−5.9 to −3.1) DBP: −3.0 (−4.8 to −1.1)	Potassium supplementation in hypertensives was generally associated with decreased blood pressure, particularly in high sodium consumers.
Poorolajal et al., 2017 [41]	23	Randomized controlled trials	1213	Primary hypertension: 732 (SBP) 695 (DBP)	4–52 weeks	6–200 mmol/day	SBP: −4.3 (−6.0 to −2.5) DBP: −2.5 (−4.1 to −1.0)	Potassium supplementation has a modest but significant impact on blood pressure.
Bommel and Cleophas 2012 [42]	10	Cross-over and parallel design studies	556	High salt intake, >170 mmol/24h	Follow up 8–16 weeks	Not available	SBP: −9.5 (−10.8 to −8.1) DBP: −6.4 (−7.3 to −5.6)	Potassium treatment reduces considerably the blood pressure of hypertensive patients on salt rich diets.

Table 5 below depicts the role of **magnesium** on hypertension (Iqbal, S., Klammer, N., & Ekmekcioglu, C., 2019: Copyright © 2019 by the authors: Licensee MDPI, Basel, Switzerland. *This article is an open access article distributed under the terms and conditions of the Creative Commons Attribution (CC BY) license http://creativecommons.org/licenses/by/4.0/):*

Key: SBP = Systolic blood pressure; DBP = Diastolic blood pressure; NA = not applicable

Author-Year	No. of Trials	Study Characteristics	No. Participants	Study Aims	Duration of Trials	Magnesium Dosage (Diet or Supplement)	Blood Pressure Lowering in mmHg or RR (95% CI)	Further Remarks/Summary
Zhang et al., 2016 [11]	27	Randomized controlled trials	Magnesium group: 822 Placebo group: 800	Effect of magnesium supplementation in normotensive and hypertensive adults (age 18–84 years).	3 weeks–6 months	Median dose of 368 mg/day (range: 238–960 mg/day)	SBP: −2 (−0.4 to −3.6) DBP: −1.8 (−0.7 to −2.8)	Magnesium supplementation at a median dose of 368 mg/day for a median duration of 3 months significantly reduced SBP and DBP. Magnesium supplementation at a dose of 300 mg/day or duration of 1 month is enough to elevate serum magnesium and reduce blood pressure. Serum magnesium was negatively associated with DBP but not SBP.

Dibaba et al., 2017 [48]	11	Randomized controlled trials	543	Assessing the pooled effect of magnesium supplementation on blood pressure in participants with preclinical or non–communicable diseases.	1 to 6 months (mean: 3.6 months)	365–450 mg/day	Standardized mean difference: SBP: −0.2 (−0.4 to −0.03) DBP: −0.3 (−0.5 to −0.03)	Magnesium supplementation lowers blood pressure in individuals with insulin resistance, prediabetes, or other noncommunicable chronic diseases.
Verma and Garg 2017 [49]	28	Randomized controlled trials	1694 (834 treatment arm, 860 placebo arm)	Evaluation the effect of magnesium supplementation on type 2 diabetes associated cardiovascular risk factors in both diabetic and nondiabetic individuals. Only four studies were carried out in hypertensive subjects.	4–24 weeks	Elemental magnesium: 300–1006 mg/day	Weighted mean difference: SBP: −3.06 (−5.51 to −0.60) DBP: −1.37 (−3.02 to 0.29)	A significant improvement was observed in SBP. Insignificant improvement or no improvement was observed in DBP
Kass et al., 2012 [50]	22	Interventional studies	1173	Assessing the effect of magnesium supplementation on blood pressure. Adults from 12 different countries were included.	3 to 24 weeks of follow-up	Elemental magnesium dosage: 120–973 mg/day	Overall effect size: SBP: 0.3 (0.2 to 0.4) DBP: 0.4 (0.3 to 0.4)	Summary of all trials show a decrease in SBP of 3–4 mmHg and DBP of 2–3 mmHg. Magnesium supplementation appears to achieve a small but clinically significant reduction in blood pressure.
Rosanoff and Plesset 2013 [51]	7	Interventional studies	135 treated hypertensive subjects	Evaluation of magnesium supplementation in hypertension. Initial SBP of the patients was >155 mmHg	6 to 17 weeks	10.5–18.5 mmol magnesium-salt/day	Mean change: SBP: −18.7 (−22.5 to −15.0) DBP: −10.9 (−13.1 to −8.7)	This uniform subset of seven studies showed a strong effect of magnesium in treated hypertensive patients.
Observational studies								

Schoenaker et al., 2014 [45]	3	Observational studies	6616 pregnant women, age range 20–40 years	Assessing the effect of dietary factors, including magnesium, on hypertensive disorders of pregnant women.	NA	Not indicated	Significantly lower mean magnesium intake of mean 7.69 mg/day for women with hypertensive disorders of pregnancy (gestational hypertension and pre-eclampsia)	Pooled results revealed statistically significantly lower mean magnesium intake for women with hypertensive disorders of pregnancy.
Han et al., 2017 [52]	10	Prospective cohort studies	180,566 participates	Assessing the relationship between dietary magnesium intake and serum magnesium concentrations on the risk of hypertension in adults. Adult population >18 years was included.	4–15 years	96–425 mg/day	RR: 0.95 (0.90 to 1.00) for a 100 mg/increment in magnesium intake. Comparing highest to lowest: RR: 0.91 (0.80 to 1.02)	Increase in magnesium intake was associated with a lower risk of hypertension in a linear dose-response pattern.
Wu J et al., 2017 [53]	3	Prospective cohort studies with four cohorts	14,876 participants (3149 cases)	Evaluation of circulating magnesium levels and incidence of coronary heart diseases, hypertension, and type 2 diabetes mellitus	Average of 6.7 years of follow-up	NA	Per 0.1 mmol/L increment in serum magnesium levels: RR: 0.96 (0.93 to 0.99)	A significant inverse linear association was observed between circulating magnesium levels and incidence of hypertension.

The *RDA: Recommended daily allowance* and *RDI: Reference daily intake* for micronutrients along with the source are represented below (Chiu, H. F., Venkatakrishnan, K., Golovinskaia, O., & Wang, C. K., 2021:

Micronutrients	**Source**	**RDA/RDI (Dietary Reference) Value/Day** *
Sodium	Table salt, processed food, seaweed, cheese	>1500 mg
Potassium	Grains, lean meat, milk, potato, banana, avocado	4700–4800 mg
Magnesium	Nuts, grains, pumpkin seed, avocado, spinach	350–420 mg
Zinc	Grains, red meat, egg, salmon, oyster, crab	8–12 mg
Calcium	Milk, tofu, okra, thyme, soybean, salmon, kale	1000–1300 mg
Selenium	Brazil nuts, seafood, organ meat, grains, egg	50–55 µg
Copper	Nuts, organ meat, oyster, grains, beans	1–1.8 mg
Vitamin B6	Milk, egg, lean meat, tuna, pistachios	1.3–1.8 mg
Vitamin B12	Liver, egg, salmon, mussels, red meat	1.5–2.4 µg
Vitamin C	Citrus fruits, guava, kiwi fruit, broccoli, pepper	70–90 mg
Vitamin D	Salmon, tuna, herring, egg, mushroom, meat	10–20 µg
Vitamin E	Spinach, avocado, nuts, seeds oil, bell pepper	10–15 mg

3.2 MIRNAS OF THE GUT-LIVER AXIS

Homeostasis entails bidirectional communication along the gut-liver axis (GLA). A slew of elements ranging from bilirubin, bile salts and lipids are produced by the liver to be stored in the gallbladder. Secretion of these constituents following food intake allows for the assimilation of nutrients in the intestine. The gut microbiota or microbes in the large intestine (refer to the book by the same authors on gut microbiota by ShivSanjeevi Sripathi and Prerna Pandey for more insights) allow for digestion of food with immune stimulation of the liver also facilitated by Kupffer cells of the liver. The nutrients from the gut to the liver also facilitate xenobiotic metabolism thus, making this link bidirectional as shown below in figure 3 (Golonka et al, 2021:*Open Access: This article is licensed under a Creative Commons Attribution 4.0 International License, which permits use, sharing, adaptation, distribution and reproduction in any medium or format, as long as you give appropriate credit to the original author(s) and the source, provide a link to the Creative Commons licence, and indicate if changes were made. The images or other third party material in this article are included in the article's Creative Commons licence, unless indicated otherwise in a credit line to the material. If material is not included in the article's Creative Commons licence and your intended use is not permitted by statutory regulation or exceeds the permitted use, you will need to obtain permission directly from the copyright holder. To view a copy of this licence, visit http://creativecommons.org/licenses/by/4.0/*):

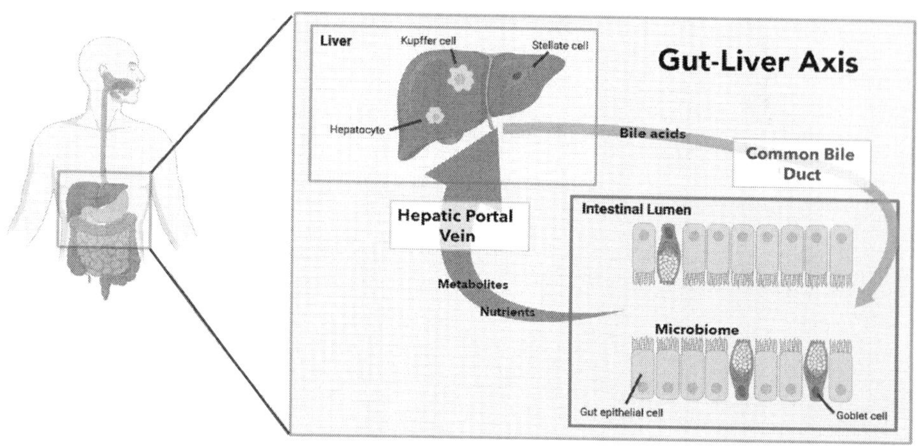

MicroRNAs (miRNAs) are described as 21–25 nucleotide long non-coding RNAs associated with a slew of processes in animals and plants. The role of these molecules in the GLA and essential hypertension (EH) is also being slowly uncovered. The levels of miRNA-615-3p were found to be high in portal hypertensive patients subjected to splenectomy. Further, EH patients have displayed a dip in the levels of miR-29a, miR-29b, miR-29c, miR-126, miR-26b, miR-9, miR-146a, etc. with an increased expression of miR-1, miR-202-3p, miR-505, miR-1183, etc. Such altered miRNA expression has also been connected with several cardiovascular diseases opening up the avenue of employing miRNAs as diagnostic and therapeutic tools for hypertension. Further, EH patients have also been observed to demonstrate lower absorption of gut microbiota-produced short-chain fatty acids (SCFAs) (Golonka et al, 2021).

Figure 4 below is illustrative of changes in miRNAs of the gut-liver axis due to certain dietary molecules such as lipids or proteins or carbohydrates via renin-angiotensin-aldosterone or altered vascularity to manifest as hypertension (Golonka et al, 2021:*Open Access: This article is licensed under a Creative Commons Attribution 4.0 International License, which permits use, sharing, adaptation, distribution and reproduction in any medium or format, as long as you give appropriate credit to the original author(s) and the source, provide a link to the Creative Commons licence, and indicate if changes were made. The images or other third party material in this article are included in the article's Creative Commons licence, unless indicated otherwise in a credit line to the material. If material is not included in the article's Creative Commons licence and your intended use is not permitted by statutory regulation or exceeds the permitted use, you will need to obtain permission directly from the copyright holder. To view a copy of this licence, visit http://creativecommons.org/licenses/by/4.0/*):

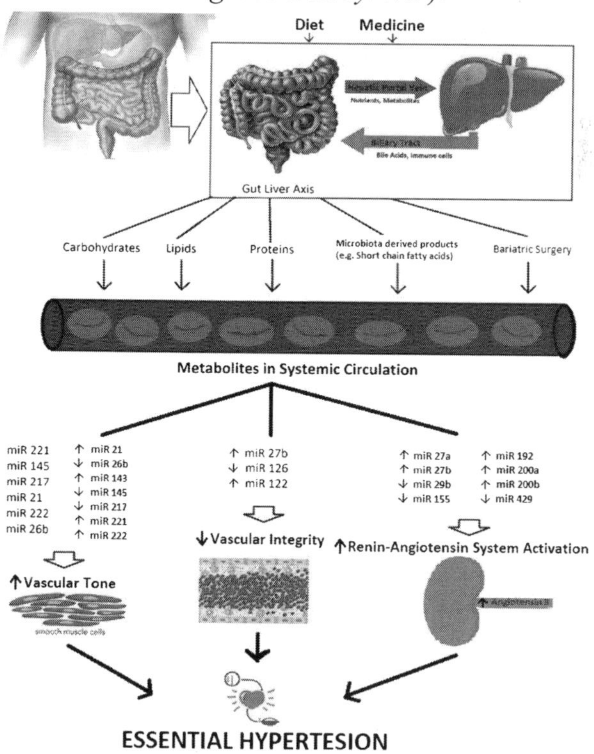

Interestingly, gut microbe dysbiosis and genetic hypertension were prevented in Wistar Kyoto rats by treatment with gut microbes *Bifidobacterium breve* CECT7263 (BFM) and *Lactobacillus fermentum* CECT5716 (LC40) (Robles-Vera et al, 2020) paving the way for gut microbe modulation for addressing hypertension.

Metabolomics

Interest in looking at the profile of metabolites (amino acids, vitamins, carbohydrates, nucleic acids, minerals, alkaloids, etc.) or metabolomics in hypertensive vs. control subjects is on the rise. Early work in 2003 by Brindle and team showed that borderline SBP (131-149 mm Hg) and high SBP (> or = 150 mm Hg) could be differentiated from low/normal systolic blood pressure (SBP < or = 130 mm Hg) employing 1H NMR spectroscopy. Other studies demonstrating differences in the metabolome due to hypertension are summarized below in Table 6 below as described in a 2020-published article in *Clinical Hypertension* (Onuh and Aliani, 2020: *Open Access: This article is licensed under a Creative Commons Attribution 4.0 International License, which permits use, sharing, adaptation, distribution and reproduction in any medium or format, as long as you give appropriate credit to the original author(s) and the source, provide a link to the Creative Commons licence, and indicate if changes were made. The images or other third party material in this article are included in the article's Creative Commons licence, unless indicated otherwise in a credit line to the material. If material is not included in the article's Creative Commons licence and your intended use is not permitted by statutory regulation or exceeds the permitted use, you will need to obtain permission directly from the copyright holder. To view a copy of this licence, visit http://creativecommons.org/licenses/by/4.0/. The Creative Commons Public Domain Dedication waiver (http://creativecommons.org/publicdomain/zero/1.0/) applies to the data made available in this article, unless otherwise stated in a credit line to the data*):

[a]*YDYHS - Yin-deficiency and Yang-hyperactivity syndrome; YYDS: Yin-Yang deficiency syndrome*

S/N	Samples	Sample size	Methodology	Results	References
1.	Spontaneously hypertensive rats (SHR) and normotensive controls, Wistar Kyoto rats (WKY)	6 SHR and WKY each (total $n=12$)	^{1}H-NMR-based urine metabolomics	There was clear separation in the principal components between the 2 samples. This was attributed to comparatively low levels of citrate, α-ketoglutarate, and hippurate, and high levels of creatine, creatinine, and some other metabolites in the urines of SHR than WKY.	[7]
2.	Spontaneously hypertensive rats (SHR) and their aged-matched Wistar Kyoto rats (WKY)	6 SHR and WKY each (total $n=12$)	^{1}H-NMR-based 24 h urine metabolomics	Major metabolic changes were observed for the rats including differences in citrate, a-ketoglutarate, succinate, hippurate, phenylacetylglycine, p-cresol glucuronide, creatine, taurine, and medium chain dicarboxylates	[23]
3.	Spontaneously hypertensive rats (SHR) and their aged-matched Wistar Kyoto rats (WKY)	29 SHR and 18 WKY	LC-QTOF-MS of urine and plasma metabolomics	Partial least squares of several metabolites found in the urine and plasma samples of rats (symmetric dimethylarginine, N2-acetyl-L-ornithine, buthionine sulfoximine, uric acid, α-tocopherol succinate, L-isoleucine, creatinine, and phospholipids) were significantly correlated with hypertension.	[26]

4.	Patients with low/normal systolic blood pressure (SBP ≤130 mmHg), borderline SBP (131–149 mmHg) and high SBP (≤150 mmHg; n=17)	SBP ≤130 mmHg, n=28; SBP (131–149 mmHg, n=19; SBP (≤150 mm; Hg, n=17; Total n=64	^{1}H-NMR of serum metabolites	The PCA showed clear differences between serum profiles of patients with low/normal SBP and that of the patients with borderline or high SBP but not between samples of borderline and high SBP. These differences were attributed to differences in serum lipid moieties between the groups.	[21]
5.	Uppsala Longitudinal Study of Adult Men Cohort	504 individuals	GC-MS of plasma	Plasma levels of ceramide, triacylglycerol, total glycerolipids and oleic acid directly correlated with longitudinal changes diastolic BP while cholesterylester levels on the other hand inversely correlated with longitudinal diastolic BP change. However, two glycerolipids validated in an independent cohort. These results suggest a pathophysiological pathophysiological pathways of hypertension	[35]
6.	Chinese Municipal Cohort Study (CMCS)	460	GC-MS of serum	241 metabolites identified, 26 were significantly different between hypertension and control at baseline and 16 out of these were associated with hypertension after adjusting for BMI, smoking and drinking.	[14]

7.	Sympathetic activity and Ambulatory Blood Pressure in Africans (SABPA)	25	GC-MS, LC-MS of urine	38 metabolites differed greatly between the two blood pressure groups with clear PCA separation. Pathway analysis showed changes in ethanol metabolism.	[36]
8.	Hypertensive and normotensive Uygur patients	250	^{1}H-NMR of plasma metabolites	OPLS-DA showed clear separation indicating differences between the two groups. Twelve different metabolites were identified suggesting biomarkers of hypertension.	[37]
9.	Case controlled study of 64 essential hypertension and 59 healthy controls	123	NMR of filtered serum	Several metabolites were identified but six of these (alanine, pyruvate, methionine, arginine, adenine, and uracil) were discovered to significantly differentiate EH from HC cohorts. Arginine was up-regulated and alanine, pyruvate, methionine, adenine, and uracil were down-regulated in EH compared to HC.	[18]
10.	European Prospective Investigation Into Cancer and Nutrition (EPIC)–Potsdam study	1116 (135 cases and 981 non cases) of incident hypertension	AbsoluteIDQ p150 Kits (Biocrates Life Scienes AG, Innsbruck, Austria) based on flow injection analysis tandem mass spectrometry technique	127 metabolites were validated to predict hypertension. Six of these were identified as the most predictive biomarkers of incident hypertension. Up-regulation of serine, glycine, and acyl-alkyl-phosphatidylcholines (C42:4 and C44:3) were associated with higher and down-regulation of diacyl-phosphatidylcholines C38:4 and C38:3 with lower predicted 10-year hypertension free survival.	[3]

11.	Black normotensives from The Atherosclerosis Risk in Communities Study	896 black normotensives including 565 women aged 45–64 years	Untargeted GC-MS and LC-MS based quantification of serum samples metabolites.	204 metabolites were measured during a 4–6 weeks period. 38% of the baseline normotensives measured developed hypertension during a 10 years follow-up period. Up-regulation of 4-hydroxyhippurate, a product of gut microbial fermentation was found to significantly increase the risk of hypertension by 17%. PCA yielded sex steroids, α - amino acids, and branch-chain amino acids with only the sex hormone revealed to be significantly associated with hypertension.	[17]
12.	Participants in The Atherosclerosis in Risk Communities Study	9104 ARIC participants without hypertension at baseline		Serum uric acid was positively correlated with incident hypertension. The association is more pronounced with Blacks than Whites.	[38]
13.	San Antonio Family Heart Study	1192 individuals drawn from 42 families	HPLC combined with tandem MS plasma lipidomics of 319 lipid species	Diacylglycerols, especially DG 16:0/22:5 and DG 16:0/22:6 were significantly associated with incident hypertension during follow-up. Four lipid species, including the DG 16:0/22:5 and DG 16:0/22:6 species were also genetically correlated with hypertension.	[39]

14.	Patients with essential hypertension and its Chinese subtypes [a](YDYHS and YYDS)	Controls (n = 22); YDYHS (n = 31); YYDS (n = 29); Total n=82	^{1}H-NMR and GC-MS of plasma metabolites	The PCA and PLS-DA showed clear separation between the groups with a little overlap between YDYHS and healthy volunteers. Glucose was found to be markedly increased in both YDYHS and YYDS groups, suggesting abnormality in glucose metabolism.	[16]

"Tomorrow is the most important thing in life. Comes into us at midnight very clean. It's perfect when it arrives and it puts itself in our hands. It hopes we've learned something from yesterday."

—John WayneR

Author's note: This chapter is not meant to be diagnostic or to be considered as a treatment option. Please consult a healthcare giver or center for relevant advice.

3.3 REFERENCES

1. Challa HJ, Ameer MA, Uppaluri KR. DASH Diet To Stop Hypertension. [Updated 2021 May 19]. In: StatPearls [Internet]. Treasure Island (FL): StatPearls Publishing; 2021 Jan-. Available from: https://www.ncbi.nlm.nih.gov/books/NBK482514/
2. Soltani, S., Arablou, T., Jayedi, A. *et al.* Adherence to the dietary approaches to stop hypertension (DASH) diet in relation to all-cause and cause-specific mortality: a systematic review and dose-response meta-analysis of prospective cohort studies. *Nutr J* 19, 37 (2020). https://doi.org/10.1186/s12937-020-00554-8
3. Chiu, H. F., Venkatakrishnan, K., Golovinskaia, O., & Wang, C. K. (2021). Impact of Micronutrients on Hypertension: Evidence from Clinical Trials with a Special Focus on Meta-Analysis. *Nutrients*,*13*(2), 588. https://doi.org/10.3390/nu13020588
4. Iqbal, S., Klammer, N., & Ekmekcioglu, C. (2019). The Effect of Electrolytes on Blood Pressure: A Brief Summary of Meta-Analyses. *Nutrients*, *11*(6), 1362. https://doi.org/10.3390/nu11061362
5. Golonka, R. M., Cooper, J. K., Issa, R., Devarasetty, P. P., Gokula, V., Busken, J., Zubcevic, J., Hill, J., Vijay-Kumar, M., Menon, B., & Joe, B. (2021). Impact of Nutritional Epigenetics in Essential Hypertension: Targeting microRNAs in the Gut-Liver Axis. *Current hypertension reports*, *23*(5), 28. https://doi.org/10.1007/s11906-021-01142-9
6. ShivSanjeevi Sripathi and Prerna Pandey (2021), Gut microbiota, Delve publications: In press
7. Robles-Vera I, Toral M, de la Visitación N, Sánchez M, Gómez-Guzmán M, Romero M, Yang T, Izquierdo-Garcia JL, Jiménez R, Ruiz-Cabello J, Guerra-Hernández E, Raizada MK, Pérez-Vizcaíno F, Duarte J. Probiotics Prevent Dysbiosis and the Rise in Blood Pressure in Genetic Hypertension: Role of Short-Chain Fatty Acids. Mol Nutr Food Res. 2020 Mar;64(6):e1900616. doi: 10.1002/mnfr.201900616. Epub 2020 Feb 6. PMID: 31953983.
8. Onuh, J.O., Aliani, M. Metabolomics profiling in hypertension and blood pressure regulation: a review. *Clin Hypertens* 26, 23 (2020). https://doi.org/10.1186/s40885-020-00157-9

9. Brindle JT, Nicholson JK, Schofield PM, Grainger DJ, Holmes E. Application of chemometrics to 1H NMR spectroscopic data to investigate a relationship between human serum metabolic profiles and hypertension. Analyst. 2003 Jan;128(1):32-6. doi: 10.1039/b209155k. PMID: 12572799.

Chapter 4

Changes in the Body Reported Due to Hypertension-Case Studies on the Brain, Depression, Quality of Life and Diabetes

Shivsanjeevi Sripathi

CONTENTS

4.1 The Brain ... 146
4.2 Hypertensive vs. Non-Hypertensive ... 149
4.3 Pre-hypertensive vs. Normotensive ... 150
4.4 Medicated vs. Non-medicated ... 151
4.5 Hypertensive vs. Non-hypertensive ... 153
4.6 Pre-hypertensive vs. Normotensive ... 153
4.7 Medicated vs. Non-medicated ... 153
4.8 The Blues: Depression ... 166
4.10 Diabetes and Hypertension ... 205
4.11 Final Points ... 215
4.12 References ... 216

4.1 THE BRAIN

An analysis of 70 to 80-year-old ladies in communities in eastern Baltimore as part of the Women's Health and Aging Study (WHAS) II scrutinized the impact of elevated baseline systolic blood pressure (SBP) on **cognition**. Along with the measurement of the blood pressure, cognition was evaluated employing the Mini-Mental State Exam (MMSE) that assesses orientation, attention, calculus, recall and language. Subsequent analyses of the assessments of 336 ladies at baseline and over nine years (mean age=74.1 years) were done on the basis of the control group (SBP≤139 mmHg), HTN I group (SBP=140–159 mmHg) and HTN II group (SBP≥160 mmHg).

In comparison with the control, the systolic blood pressure, diastolic blood pressure and pulse pressures in the HTN II group participants were higher. On the basis of Trail Making Part A (TMT-A) scores of cognitive decline, the odds of impairment were lower for HTN I participants, as reported by the cross-sectional analyses. Further, stratification on the basis of age revealed that the risk of impairment in 76-80 year-olds on Trail Making Part B (TMT-B) scores was higher in the HTN II group members vs. controls by a factor of five. Also, the risk of cognitive impairment was more in 70-75-year-olds by a factor of three or four when the PP (pulse pressure) exceeded 84 mmHg vs. controls.

(Yasar et al, 2011)

Table 1 below depicts ***TOP***: the features of the 336 participants (© 2011 Yasar et al. *This is an open-access article distributed under the terms of the Creative Commons Attribution License, which permits unrestricted use, distribution, and reproduction in any medium, provided the original author and source are credited*):

Characteristics					PP	PP	PP	Statistics
	Control Group	**HTN I**	**HTN II**	**Statistics**	**Lower Tertile**	**Middle Tertile**	**Upper Tertile**	
	(N = 103)	**(N = 124)**	**(N = 109)**	*p*	**(N = 117)**	**(N = 107)**	**(N = 112)**	*p*
	N (%)				**N (%)**			
Age (Mean, SD)	73.5 (2.9)	73.7 (2.8)	74.1 (2.7)	0.2	73.4 (2.7)	73.9 (2.9)	74.02 (2.7)	0.2
Race:								
White	87 (84.5)	101 (81.5)	97 (89.0)	0.3	90 (76.9)	95 (88.8)	99 (88.4)	0.02
Education:								
<12 years	9 (8.7)	11 (8.9)	7 (6.4)	0.5	12 (10.3)	12 (11.2)	3 (2.7)	0.01
= 12 years	43 (41.7)	61 (49.2)	57 (52.3)		48 (41.0)	45 (42.1)	68 (60.7)	
>12 years	51 (49.5)	52 (41.9)	45 (41.3)		57 (48.7)	50 (46.7)	41 (36.6)	
Income ($/year):								
<10,000	20 (24.4)	25 (24.5)	21 (23.1)	0.5	24 (24.5)	24 (27.9)	18 (20.0)	0.6
10,000–24,999	27 (32.9)	32 (31.4)	38 (41.8)		33 (33.7)	26 (30.2)	38 (42.2)	
25,000–49,999	19 (23.2)	23 (22.6)	22 (24.2)		26 (26.5)	20 (23.3)	18 (20.0)	
>50,000	16 (19.5)	22 (21.6)	10 (11.0)		15 (15.3)	16 (18.6)	16 (17.8)	
Marital status:								
Married	64 (62.1)	70 (56.5)	70 (64.2)	0.4	77 (65.8)	60 (56.1)	68 (60.7)	0.3
Not Married	39 (37.9)	54 (43.6)	39 (35.8)		40 (34.2)	47 (43.9)	44 (39.3)	
Smoking:								
Never	57 (55.3)	67 (54.0)	57 (52.8)	0.91	65 (55.6)	59 (55.1)	58 (52.3)	0.9
Ever	46 (44.7)	57 (46.0)	51 (47.2)		52 (44.4)	48 (44.9)	53 (47.8)	
Hx HTN	29 (27.9)	68 (54.8)	75 (68.8)	<0.001	51 (43.6)	52 (48.6)	69 (61.6)	0.02
Hx Stroke	4 (3.9)	4 (3.2)	4 (3.7)	0.7	5 (4.3)	4 (3.7)	3 (2.7)	0.5
Hx MI	6 (5.8.2)	11 (8.9)	11(10.1)	0.5	5 (4.3)	7 (6.5)	16 (14.3)	0.02
Hx Angina	19 (8.3)	18 (14.5)	14 (12.8)	0.5	16 (13.7)	19 (17.6)	16 (14.3)	0.7
Hx CHF	6 (5.8)	8 (6.5)	9 (8.3)	0.8	10 (8.6)	2 (1.9)	11 (9.8)	0.04
Hx PAD	10 (9.6)	9 (7.3)	12 (11.0)	0.6	12 (10.3)	6 (5.6)	13 (11.6)	0.3
Hx DM	8 (7.7)	12 (9.7)	6 (5.5)	0.5	11 (9.4)	7 (6.5)	8 (7.1)	0.7
Ever taking HTN medications								
	54 (51.9)	101 (81.5)	92 (84.4)	<0.001	77 (65.8)	79 (73.8)	90 (80.4)	0.04
	Mean (SD)				**Mean (SD)**			
SBP (mmHg)	129.1 (9.1)	148.9 (5.0)	177.6 (12.0)	<0.001	138.1 (18.4)	150.6 (17.1)	167.9 (17.1)	<0.001
DBP (mmHg)	67.4 (10.1)	73.4 (11.8)	88.2 (17.0)	<0.001	80.0 (15.9)	76.6 (16.8)	72.1 (13.5)	<0.001
PP (mmHg)	61.7 (12.6)	75.5 (12.7)	89.5 (17.7)	<0.001	58.1 (10.0)	73.9 (3.4)	95.8 (12.0)	<0.001
BMI	25.9 (5.7)	27.3 (5.0)	26.5 (4.6)	0.1	26.7 (5.4)	26.2 (4.5)	27.0 (5.4)	0.5
GDS	3.7 (3.6)	3.9 (4.1)	3.5 (2.9)	0.7	3.4 (3.3)	4.2 (4.0)	3.7 (3.5)	0.3
Mean years to follow up								
	6.4 (8.7)	7.0 (6.0)	6.6 (6.2)	0.8	6.4 (8.2)	7.4 (4.9)	6.3 (7.3)	0.4

Control group (SBP≤139 mmHg); HTN I = Hypertension I (SBP 140–159 mmHg); HTN II = Hypertension II (SBP≥160); PP lower tertile = Pulse pressure 48–68 mmHg, PP middle tertile = Pulse pressure 71–77 mmHg, and PP upper tertile = Pulse pressure 84–108 mmHg.
Statistics = Chi-square test was used for categorical and ANOVA with pairwise comparisons for continuous variables.
Hx HTN = history of high blood pressure, Hx stroke = history of stroke, Hx MI = history of myocardial infarction, Hx Angina = history of angina, Hx CHF = history of congestive heart failure, Hx PAD = history of peripheral artery disease, Hx DM = history of diabetes mellitus; BMI = Body Mass Index, GDS = Geriatric Depression Scale.
doi:10.1371/journal.pone.0027976.t001

BOTTOM: the longitudinal analyses of the effects of elevated blood pressure and cognitive decline in 70 to 80 year-old ladies (non-demented) over nine years ():

	HTN I	HTN II	PP Middle Tertile	PP Upper Tertile
Ages 70–75 (n = 248)				
	HR (95% CI)	HR (95% CI)	HR (95% CI)	HR (95% CI)
HVLT-D				
Unadjusted	1.46 (0.74–2.87)	1.05 (0.49–2.27)	1.06 (0.54–2.07)	1.08 (0.55–2.11)
Model 1	1.40 (0.71–2.76)	0.91 (0.42–1.99)	1.15 (0.58–2.31)	1.25 (0.61–2.60)
Model 2	1.39 (0.68–2.84)	0.88 (0.39–2.02)	0.99 (0.47–2.09)	1.32 (0.63–2.78)
HVLT-I				
Unadjusted	1.57 (0.78–3.13)	1.48 (0.70–3.14)	1.92 (0.96–3.87)	1.77 (0.87–3.58)
Model 1	1.52 (0.76–3.03)	1.28 (0.60–2.72)	2.58 (1.24–5.36)**	2.65 (1.21–5.80)*
Model 2	1.47 (0.71–3.02)	1.21 (0.55–2.66)	2.44 (1.11–5.39)*	3.13 (1.35–7.24)**
TMT, Part A				
Unadjusted	2.52 (1.07–5.89)*	1.25 (0.45–3.46)	0.78 (0.35–1.73)	0.93 (0.43–1.99)
Model 1	2.38 (1.01–5.57)*	1.08 (0.39–3.00)	1.06 (0.47–2.39)	1.86 (0.81–4.26)
Model 2	2.13 (0.84–5.42)	1.02 (0.35–3.00)	1.01 (0.38–2.68)	1.46 (0.57–3.74)
TMT, Part B				
Unadjusted	1.15 (0.61–2.16)	1.05 (0.53–2.07)	0.64 (0.31–1.34)	1.71 (0.95–3.07)
Model 1	1.09 (0.58–2.07)	0.87 (0.43–1.74)	0.68 (0.32–1.45)	1.83 (0.96–3.49)
Model 2	1.01 (0.51–1.99)	1.04 (0.50–2.16)	0.58 (0.25–1.32)	1.83 (0.91–3.71)
MMSE				
Unadjusted	0.65 (0.25–1.70)	1.43 (0.60–3.41)	0.97 (0.40–2.33)	0.98 (0.41–2.37)
Model 1	0.56 (0.22–1.47)	1.02 (0.43–2.46)	1.02 (0.42–2.49)	1.00 (0.38–2.60)
Model 2	0.47 (0.17–1.30)	1.07 (0.41–2.80)	1.03 (0.39–2.72)	0.97 (0.34–2.73)
Ages 76–80 (n = 88)				
	HTN I	**HTN II**	**PP Middle Tertile**	**PP Upper Tertile**
	HR (95% CI)	HR (95% CI)	HR (95% CI)	HR (95% CI)
HVLT-D				
Unadjusted	0.93 (0.39–2.25)	0.88 (0.39–2.00)	0.65 (0.27–1.55)	0.79 (0.34–1.83)
Model 1	0.85 (0.35–2.08)	0.92 (0.40–2.08)	0.64 (0.26–1.56)	0.79 (0.33–1.90)
Model 2	0.82 (0.27–2.47)	1.13 (0.38–3.43)	0.80 (0.27–2.31)	2.08 (0.54–7.98)
HVLT-I				
Unadjusted	0.85 (0.38–1.91)	0.59 (0.27–1.30)	0.82 (0.37–1.81)	0.51 (0.21–1.20)
Model 1	0.81 (0.36–1.84)	0.61 (0.28–1.35)	0.80 (0.36–1.79)	0.47 (0.19–1.16)
Model 2	0.96 (0.38–2.45)	0.98 (0.37–2.59)	0.68 (0.27–1.71)	0.86 (0.25–2.93)
TMT, Part A				
Unadjusted	0.68 (0.24–1.93)	1.15 (0.49–2.69)	0.77 (0.29–2.06)	1.03 (0.40–2.61)
Model 1	0.65 (0.23–1.85)	1.18 (0.50–2.76)	0.81 (0.29–2.24)	1.08 (0.40–2.94)
Model 2	0.80 (0.22–2.97)	2.56 (0.73–9.06)	0.52 (0.15–1.87)	2.01 (0.49–8.28)
TMT, Part B				
Unadjusted	1.26 (0.52–3.03)	1.30 (0.59–2.88)	0.75 (0.31–1.79)	0.95 (0.41–2.18)
Model 1	1.20 (0.50–2.89)	1.39 (0.62–3.11)	0.74 (0.30–1.81)	1.05 (0.42–2.65)
Model 2	2.13 (0.65–6.95)	5.05 (1.42–18.04)*	1.24 (0.36–4.28)	5.12 (1.11–23.62)*
MMSE				
Unadjusted	-#	-#	0.74 (0.23–2.44)	1.37 (0.47–4.02)
Model 1	-#	-#	1.11 (0.30–4.09)	1.89 (0.58–6.19)
Model 2	-#	-#	0.83 (0.20–3.47)	2.36 (0.40–13.83)

Reference group for HTN: Control group (SBP≤139 mmHg): for PP: pulse pressure lower tertile (48–68 mmHg).
HTN Stage I = Hypertension Stage I (SBP 140–159 mmHg); HTN Stage II = Hypertension Satge II (SBP≥160); PP middle tertile = Pulse pressure 71–77 mmHg, and PP upper tertile = Pulse pressure 84–108 mmHg.
Model 1: age, race, education.
Model 2: age, race, education, history of high blood pressure, stroke, myocardial infarction, congestive heart failure, peripheral artery disease, diabetes mellitus, angina; history of smoking, BMI, ever on hypertensive medication, glucose, cholesterol, and depression score.
Note: All Trails B models adjust for Trails A performance.
MMSE = Mini Mental State Exam; TMT, Part A = Trail Making Test, Part A; TMT, Part B = Trail Making Test, Part B; HVLT = Hopkins Verbal Learning Test (HVLT-I = immediate recall; HVLT-D = delayed recall);
*p-value<0.05.
**p-value<0.01.
doi:10.1371/journal.pone.0027976.t005

In a research article published in 2017, Suzuki and team analyzed the variations in the microstructure of the **white matter of the brain**. The approach entailed the analyses of diffusion tensor imaging (DTI) and neurite orientation dispersion and density imaging (NODDI) data from 4939 UK Biobank (UKB) participants (hypertensive vs. prehypertensive vs. normal). The processing of images entailed white matter tracing surface meshes of 27 anatomically defined white matter tracts along with measures of brain volume measures viz. total brain volume (TBV), grey matter volume (GMV) and white matter volume (WMV). Following exclusion criteria, 4659 was the final sample number. Table 2 below depicts the details of these participants (© 2017 Suzuki et al. *This is an open access article distributed under the terms of the Creative Commons Attribution License, which permits unrestricted use, distribution, and reproduction in any medium, provided the original author and source are credited*):

	Non-hypertensive (N = 2322)		Hypertensive (N = 2337)	
	Normotensive	Pre-hypertensive	Unmedicated	Medicated
N	741	1581	1290	1047
Baseline characteristics				
Age (yrs)	59.0±7.1	61.0±7.5	63.1±7.0	65.6±6.2
Male (%)	21.6	46.4	53.3	58.2
Educational level (%)	52.0	47.9	41.9	41.1
Body mass index	24.5±3.5	26.1±4.0	27.0±4.4	28.5±4.6
Ever smoked (%)	52.0	53.4	53.9	56.4
Diabetes mellitus (%)	1.8	2.5	3.0	11.7
Cardiovascular disease (%)	1.2	1.7	2.3	9.3
Blood pressures and antihypertensive medication				
Systolic blood pressure (mmHg)	111.8±6.4	129.6±5.8	152.3±11.8	142.6±17.0
Diastolic blood pressure (mmHg)	68.5±6.0	76.5±6.4	86.2±9.1	81.3±9.6
Antihypertensive medication (%)	0.0	0.0	0.0	100.0

Continuous variables were expressed as means ± standard deviations.

https://doi.org/10.1371/journal.pone.0187600.t001

The contrast between the microstructures was represented in an illustrative format across the various groups scrutinized as shown below in Figure 1 (© 2017 Suzuki et al. *This is an open access article distributed under the terms of the Creative Commons Attribution License, which permits unrestricted use, distribution, and reproduction in any medium, provided the original author and source are credited*):

Key: FA, fractional anisotropy; ICVF, intracellular volume fraction; ISOVF, isotropic volume fraction; MD, mean diffusivity

4.2 HYPERTENSIVE VS. NON-HYPERTENSIVE

It was observed that the hypertensive group displayed higher MD and higher ISOVF in the bilateral inferior fronto-occipital fasciculus, the right cingulate gyrus part of cingulum, the bilateral superior and inferior longitudinal fasciculus, the bilateral three thalamic radiations, and the forceps minor.

The MD was more in the cingulum bundle while the ISOVF was elevated in the right uncinate fasciculus. This group also presented a higher FA in the left inferior fronto-occipital fasciculus, the bilateral superior and right inferior longitudinal fasciculus, the bilateral anterior and posterior thalamic radiation, and the forceps minor. The ICVF was more bilateral superior longitudinal fasciculus and the left anterior and bilateral superior thalamic radiations.

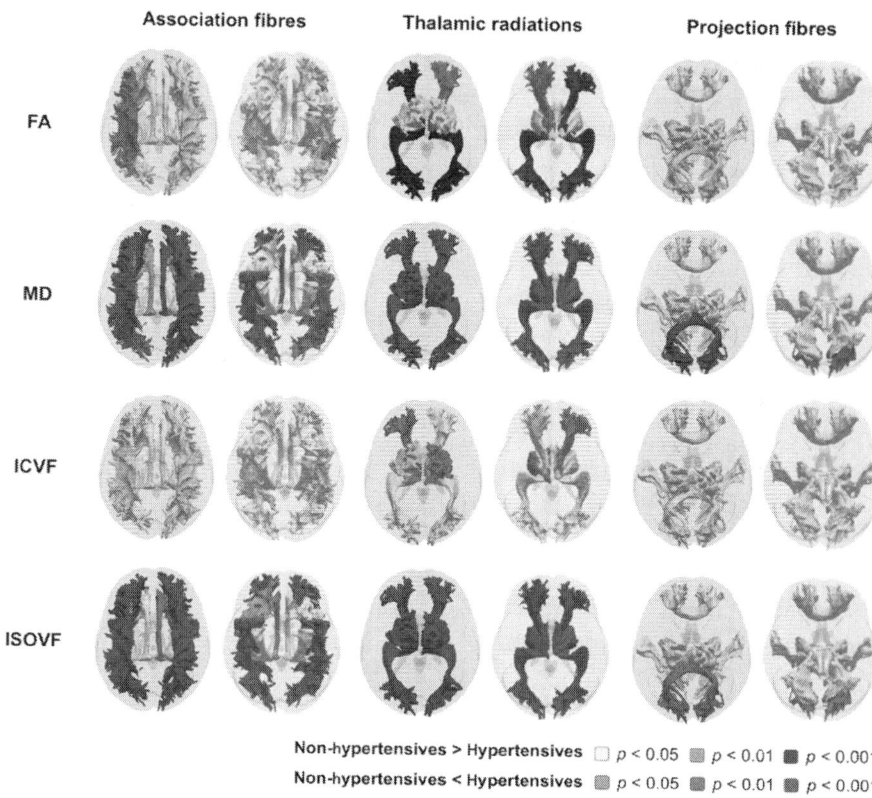

4.3 PRE-HYPERTENSIVE VS. NORMOTENSIVE

Here, a higher ISOVF was observed in the right superior longitudinal fasciculus and the right superior thalamic radiation.

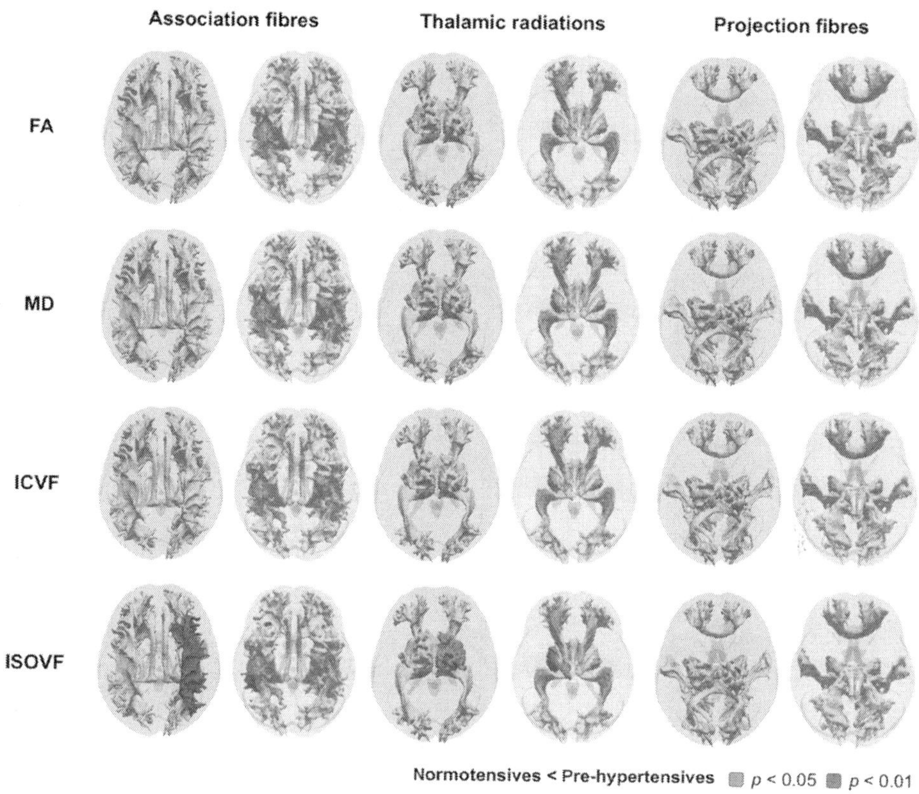

4.4 MEDICATED VS. NON-MEDICATED

An analysis of changes of hypertensive participants who were on anti-hypertensive medications vs. those hypertensive individuals who were not on these drugs revealed a dip in the FA while the MD and ISOVF were higher in those hypertensive patients on medication. The FA was lower in the bilateral superior longitudinal fasciculus, the right anterior and left posterior thalamic radiation, and the forceps minor while the MD was elevated in the bilateral superior and left inferior longitudinal fasciculus, the left anterior thalamic radiation, and the left acoustic radiation with the left superior longitudinal fasciculus, the left anterior thalamic radiation, and the left corticospinal tract displaying higher MD.

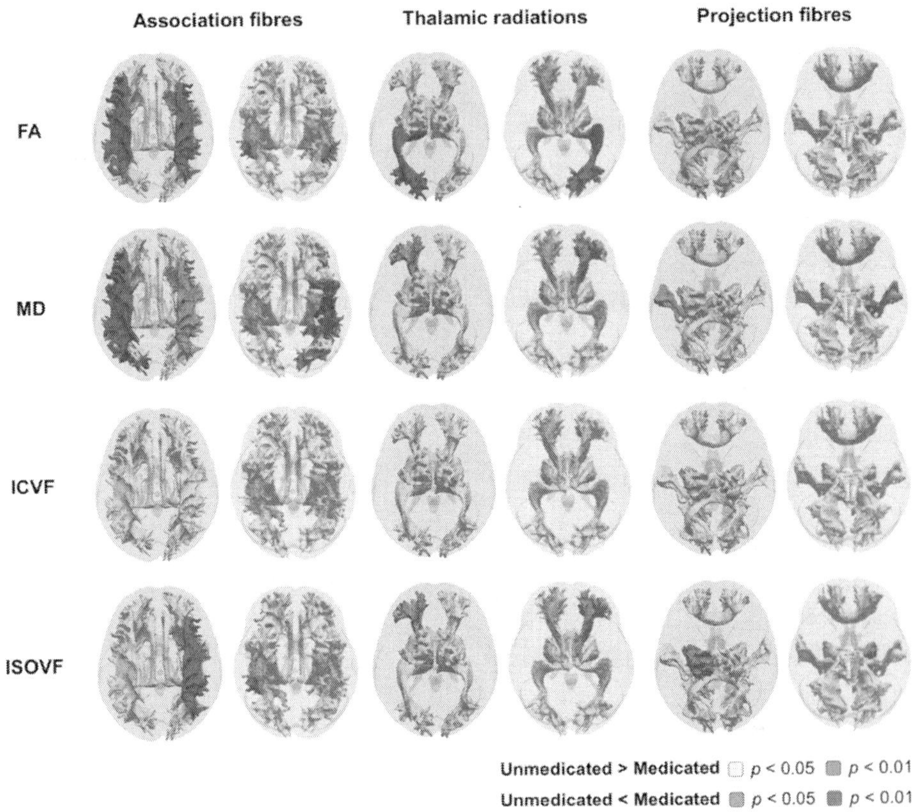

As far as the values of differences are concerned, Table 3 A to C below are representative of the trends to enable the readers to get an understanding (© 2017 Suzuki et al. *This is an open access article distributed under the terms of the Creative Commons Attribution License, which permits unrestricted use, distribution, and reproduction in any medium, provided the original author and source are credited*):

4.5 HYPERTENSIVE VS. NON-HYPERTENSIVE

	Non-hypertensive	Hypertensive	p value
N	1559	1559	
Baseline characteristics			
Age (yrs)	62.6±7.0	62.6±6.8	0.815
Male (%)	48.0	47.7	0.858
Educational level (%)	45.5	44.6	0.614
Body mass index	26.5±4.0	26.5±3.9	0.826
Ever smoked (%)	53.8	54.1	0.886
Diabetes mellitus (%)	2.9	3.5	0.358
Cardiovascular disease (%)	2.2	2.6	0.559
Blood pressures and antihypertensive medication			
Systolic blood pressure (mmHg)	125.6±9.5	148.0±14.7	<0.001
Diastolic blood pressure (mmHg)	74.5±7.1	84.5±9.4	<0.001
Antihypertensive medication (%)	0.0	38.7	<0.001

The non-hypertensive group includes normotensive and pre-hypertensive individuals. Variables were expressed as means ± standard deviations and were analyzed using the Student's t-test. Nominal variables were analyzed using the Pearson's chi-square statistic.

https://doi.org/10.1371/journal.pone.0187600.t002

4.6 PRE-HYPERTENSIVE VS. NORMOTENSIVE

	Non-hypertensive	Hypertensive	p value
N	694	694	
White matter microstructure measure			
FA	0.048±0.586	0.029±0.604	1.000
MD	–0.171±0.579	–0.127±0.591	0.949
MO	–0.033±0.337	–0.033±0.331	1.000
ICVF	0.066±0.751	0.072±0.762	1.000
OD	0.093±0.473	0.103±0.495	1.000
ISOVF	–0.178±0.479	–0.097±0.517	0.015
Brain volume measure			
Total brain volume (ml)	1531.2±72.4	1526.9±71.6	0.774
Grey matter volume (ml)	817.5±45.3	815.8±45.7	1.000
White matter volume (ml)	713.8±40.5	711.1±39.9	0.629

Variables were expressed as means ± standard deviations and were analyzed using the Student's t-test. White matter microstructural measures are expressed as mean z-scores referenced to the total study population (n = 4659). Significance levels were set at p<0.05 after Bonferroni correction. Abbreviations: FA, Fractional anisotropy; ICVF, Intracellular volume fraction; ISOVF, Isotropic volume fraction; MD, Mean diffusivity; MO, Diffusion tensor mode; OD, Orientation dispersion.

https://doi.org/10.1371/journal.pone.0187600.t005

4.7 MEDICATED VS. NON-MEDICATED

	Unmedicated	Medicated	p value
N	834	834	
White matter microstructure measure			
FA	0.028±0.619	–0.102±0.691	0.002
MD	0.052±0.651	0.191±0.708	0.001
MO	0.013±0.352	–0.003±0.344	1.000
ICVF	–0.008±0.826	–0.116±0.875	0.123
OD	–0.059±0.501	–0.055±0.507	1.000
ISOVF	0.059±0.550	0.170±0.595	0.002
Brain volume measure			
Total brain volume (ml)	1496.2±68.8	1492.1±69.2	0.806
Grey matter volume (ml)	788.8±43.8	784.8±43.6	0.267
White matter volume (ml)	707.4±40.2	707.3±41.3	1.000

Variables were expressed as means ± standard deviations and were analyzed using the Student's t-test. White matter microstructural measures are expressed as mean z-scores referenced to the total study population (n = 4659). Significance levels were set at p<0.05 after Bonferroni correction. Abbreviations: FA, Fractional anisotropy; ICVF, Intracellular volume fraction; ISOVF, Isotropic volume fraction; MD, Mean diffusivity; MO, Diffusion tensor mode; OD, Orientation dispersion.

https://doi.org/10.1371/journal.pone.0187600.t007

These findings were reported to be suggestive of axonal loss or dystrophy on account of the variations observed in the ICVF and the ISOVF in the white matter in hypertension. Further, pre-hypertension displayed an association with lower ISOVF. Elevated pressure was found to mainly target the superior longitudinal fasciculus and the superior thalamic radiation probably showing the impact of not only hypertension but also pre-hypertension on the white matter and cognition.

Another study from the Wakayama Medical University Hospital explored the changes in the functional connectivity in women with pre-eclampsia vs. women with a healthy pregnancy employing resting-state functional magnetic resonance imaging (fMRI). The diagnosis of women with pre-eclampsia (PE: n=21) vs. healthy control (HC) pregnant women (n=16) was hypertension along with proteinuria. When the systolic BP $\geq$160 mmHg or diastolic BP of $\geq$110 mmHg at least 4 hours apart on resting, it was categorized as severe PE. In the case of new detection of high BP following 20 weeks of gestation minus proteinuria, the categorization was gestational hypertension (GH) who were included in the HC group.

The analyses of diastolic BP (dBP) and regional global connectivity (rGC) as obtained from MRI analyses were first done. This value was chosen given its stability over the systolic BP and the association of dBP with complications in young ladies. A negative association was revealed between the rGC and dBP in the left orbitofrontal gyrus and right orbitofrontal gyrus (both are frontal lobe regions) as displayed below in figure 2 : ***TOP*: A)** left orbitofrontal gyrus and B) right orbitofrontal gyrus (© 2018 Kurosaki et al: *This is an open access article distributed under the terms of the Creative Commons Attribution License, which permits unrestricted use, distribution, and reproduction in any medium, provided the original author and source are credited*):

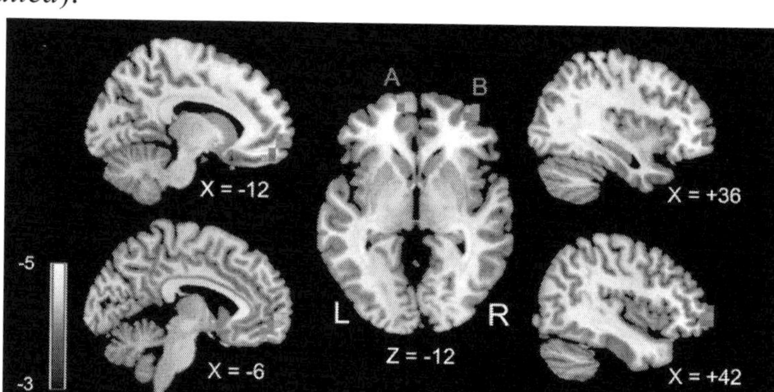

***Bottom:* the relationship between the rGC and dBP showing an inverse association in** A: left medial orbitofrontal area; B: right middle orbitofrontal area (© 2018 Kurosaki et al: *This is an open access article distributed under the terms of the Creative Commons Attribution License, which permits unrestricted use, distribution, and reproduction in any medium, provided the original author and source are credited*):

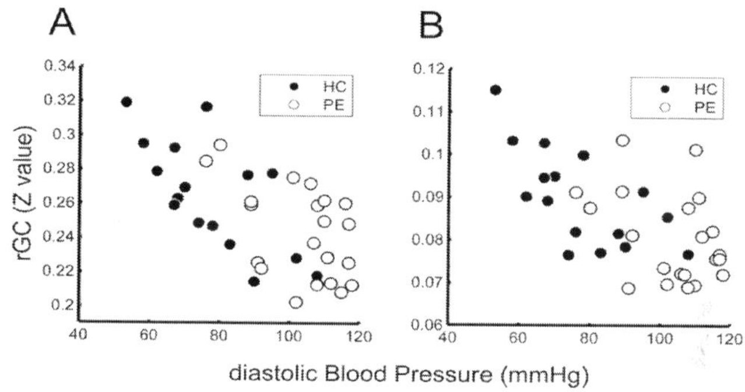

There are spontaneous neurovascular coupling interactions among the brain regions via a functional network to allow them to work together. Gaining perspectives into the effective connectivity (EC) that is the effect one neural network has on another and the functional connectivity (FC) that looks at the strength of connections between brain regions allow for understanding the impact of hypertension on this network. Researchers Bu and team (2018) employed the wavelet phase coherence (WPCO) and the coupling strength (CS) of near-infrared spectroscopy (NIRS) signals to quantify the FC and EC, respectively in hypertensive individuals vs. controls.

The study entailed the continuous recording of NIRS signals from the prefrontal cortex, sensorimotor cortex, and occipital lobes of 13 hypertensive patients (mean age= 70 ± 6.5 years old) and 16 elderly healthy subjects or the controls (mean age= 71 ± 5.5 years old). This was done in both resting and standing positions (10 minutes each) to facilitate the elucidation of neurovascular coupling functions.

The oscillations from different signals from NIRS were at different frequency intervals associated with different measurements viz. I: 0.6–2 (cardiac activity); II: 0.145–0.6 (respiratory activity); III: 0.052–0.145 (myogenic activity); and IV: 0.021–0.052 (neurogenic activity). WPCO and CS were lower at interval III and interval IV conspicuously in the participants with hypertension vs. the controls in both the resting and standing positions. Further, the hypertension group also displayed an evident decrease in WPCO and CS in the resting state as opposed to the standing state. This lowering of WPCO and CS was indicative of a lowered network interaction that was suggestive of a **disturbance in the neurovascular coupling in those with hypertension**.

Figure 3 below is illustrative of the lowering in the functional network analyses of the brains: ***TOP***: WPCO changes in (A) hypertension and control groups at rest in interval III (B) hypertension and control groups at rest in interval IV (C) hypertension and control groups at stand in interval III.: no significant differences in WPCO values (black lines), significance of the WPCO value (dark blue), a strong significance of the WPCO value (blue lines) (© 2018 Bu, Huo, Xu, Liu, Li, Fan and Li. *This is an open-access article distributed under the terms of the* ***Creative Commons Attribution License (CC BY)****. The use, distribution or reproduction in other forums is permitted, provided the original author(s) and the copyright owner are credited and that the original publication in this journal is cited, in accordance with accepted academic practice. No use, distribution or reproduction is permitted which does not comply with these terms*):

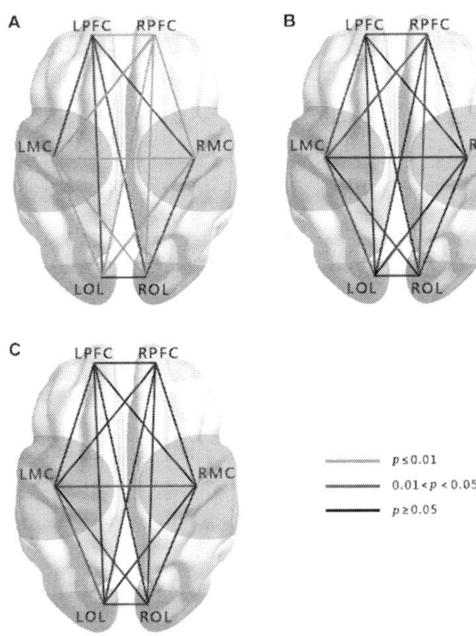

BOTTOM: Analysis of the functional connectivity (FC) at intervals III (0.052–0.145 Hz) and IV (0.021–0.052 Hz) in (A) Hypertension group at rest; (B) control group at rest; (C) hypertension group standing; and (D) control group standing (© 2018 Bu, Huo, Xu, Liu, Li, Fan and Li. *This is an open-access article distributed under the terms of the* ***Creative Commons Attribution License (CC BY)****. The use, distribution or reproduction in*

other forums is permitted, provided the original author(s) and the copyright owner are credited and that the original publication in this journal is cited, in accordance with accepted academic practice. No use, distribution or reproduction is permitted which does not comply with these terms):

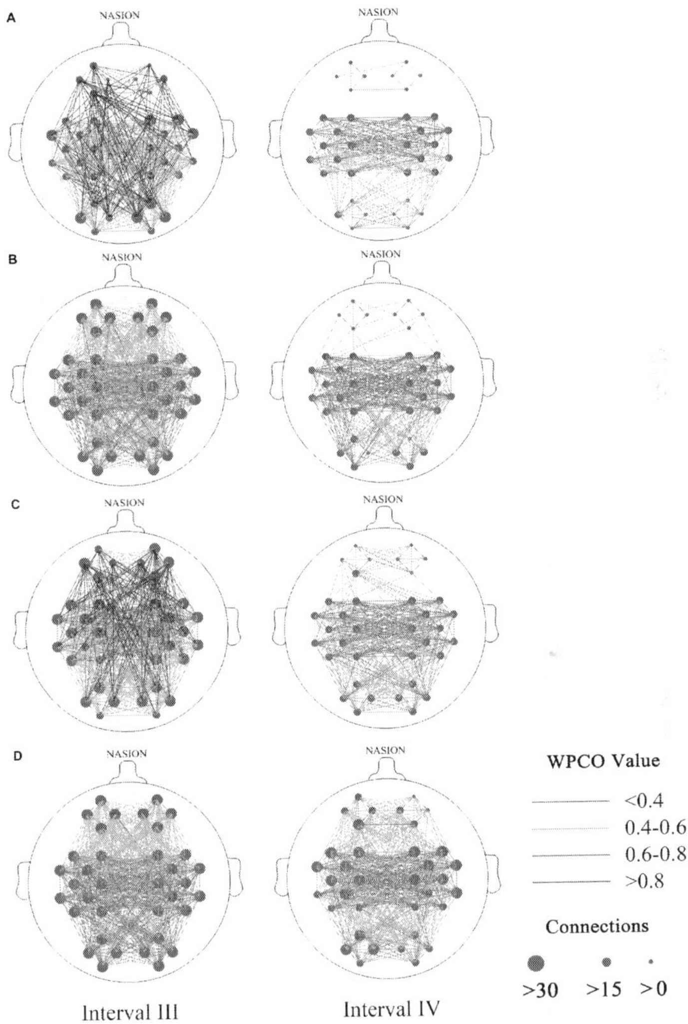

Further, evaluation of cognition as per the Mini-Mental State Examination (MMSE) (cognitive function test) and Montreal Cognitive Assessment (MoCA) scores (for attention and concentration) revealed a conspicuous lowering of cognition in those with hypertension. A positive correlation emerged between the MMSE and MoCA scores and WPCO and CS to show the association between cognition and network parameters. Given that neuronal activation can modulate neurovascular coupling to augment the blood flow in the cerebrum, aberrant neurovascular coupling in hypertension can lead to dysfunction of the brain. The researchers concluded that the results were possibly indicative of decreased connectivity across the brain regions serving to lead to weakened cognitive functioning in hypertension.

"Men and women are like right and left hands; it doesn't make sense not to use both"

-Jeannette Rankin

The variation of cognition with the blood pressure was assessed over an eight-year period by researchers Hestad, Engedal, Schirmer and Strand along with the impact of age and sex. The study participants (n=4,465) were from the longitudinal Tromsø Study Wave 6 (baseline or T0) and Wave 7 (follow-up or T1). Table 4 below depicts the details of the study (© 2020 Hestad, Engedal, Schirmer and Strand. *This is an open-access article distributed under the terms of the* ***Creative Commons Attribution License (CC BY)****. The use, distribution or reproduction in other forums is permitted, provided the original author(s) and the copyright owner(s) are credited and that the original publication in this journal is cited, in accordance with accepted academic practice. No use, distribution or reproduction is permitted which does not comply with these terms*):

	N yes	N no		Baseline characteristics at Tromsø Wave 6 (T0)					
			Age, mean (SD)	Women, (%)	Education, (% high)	Poor self-rated health (%)	SBP, mmHg (SD)	BMI, kg/m² (SD)	Digit Symbol, mean (SD)**
0. Invited to baseline examination, 45–74 years, T0	12,054								
1a. Did not participate in T0		3,241							
1b. Participated in T0	8,813		59.9 (7.8)	52	36	35	137.7 (21.9)	27.1 (4.3)	40.3 (12.6)

2a. No valid BP measurement in T0		34							
2b. Valid BP measurements in T0	8,779		59.9 (7.8)	52	36	35	137.7 (21.9)	27.1 (4.2)	40.3 (12.6)
3a. Not invited, died or emigrated before T1		416	65.1 (6.4)	40	30	56	143.3 (24.1)	26.9 (4.3)	33.0 (11.9)
3b. Invited to T1, Phase I	8,363		59.7 (7.8)	52	36	34	137.4 (21.7)	27.1 (4.2)	40.7 (12.5)
4a. Did not participate in T1, Phase I		1,895	61.0 (8.0)	52	31	42	139.9 (22.7)	27.3 (4.5)	37.3 (12.3)
4b. Participated in T1, Phase I	6,468		59.3 (7.7)	53	37	31	136.7 (21.4)	27.0 (4.2)	41.5 (12.4)
5a. Not invited to T1, Phase II (cognition)		1,172	53.5 (8.5)	42	45	28	132.1 (20.7)	27.0 (4.3)	38.3 (9.8)
5b. Invited to T1, Phase II (cognition)	5,296		60.6 (6.9)	55	36	32	137.7 (21.4)	27.0 (4.1)	41.5 (12.4)
6a. Did not participate in T1 Phase II		783	61.8 (7.3)	53	32	37	140.0 (22.8)	26.9 (4.2)	39.0 (13.0)
6b. Participated in T1 Phase II	4,513		60.4 (6.8)	55	37	31	137.3 (21.1)	27.0 (4.1)	41.9 (12.3)

7a. Died before T1 data collection was finished		17							
7b. Alive when T1 data collection was finished	4,496		60.3 (6.8)	55	37	31	137.2 (21.1)	27.0 (4.1)	41.9 (12.3)
8a. T0 BP measurements out of range*		31							
8b. T0 BP measurements within range	4,465								
9. Final Study Sample	4,465		60.3 (6.8)	55	37	31	136.8 (20.3)	27.0 (4.1)	42.0 (12.3)

* SBP > 200 mmHg, SBP < 80 mmHg, DBP > 120 mmHg, DBP < 30 mmHg. ** Available only for a subsample of 53%.

An increase in SBP was observed with increasing age at T0 and T1. This change was steeper in women compared to men with the SBP of women lower than that of men (<65 years) and higher SBP in ladies vs. men with age (65–74 years). Longitudinal analyses revealed a trend of lowered SBP in men from T0 to T1 while the 45–54-year-old women group showed a significant increase with time. Table 5 below represents these changes (

with accepted academic practice. No use, distribution or reproduction is permitted which does not comply with these terms):

Birth cohort	Age T0 (2007/2008)	Age T1 (2015/2016)	N	T0	T1	Change from T0 to T1
Men				**Systolic blood pressure**		
1953–1962	45–54	53–62	420	131.4 (16.2)	131.9 (16.9)	0.5 (16.4)
1943–1952	55–64	63–72	1,001	138.2 (18.6)	136.3 (18.2)	−1.8(18.3)
1933–1942	65–74	73–82	577	143.7 (19.3)	140.7 (18.8)	−3.0(20.1)
Wom-en						
1953–1962	45–54	53–62	585	123.9 (18.2)	125.9 (17.7)	2.1 (15.9)
1943–1952	55–64	63–72	1,217	135.5 (19.9)	136.5 (19.9)	1.0 (19.8)
1933–1942	65–74	73–82	665	145.7 (21.2)	146.2 (21.3)	0.5 (23.3)
Men				**Diastolic blood pressure**		
1953–1962	45–54	53–62	420	82.2 (9.7)	79.4 (8.9)	−2.8(9.1)
1943–1952	55–64	63–72	1,001	82.6 (10.1)	77.6 (9.7)	−5.0(9.8)
1933–1942	65–74	73–82	577	80.7 (9.4)	75.7 (9.4)	−5.0(10.1)
Wom-en						
1953–1962	45–54	53–62	585	75.0 (9.8)	73.9 (9.4)	−1.1(8.6)
1943–1952	55–64	63–72	1,217	75.9 (9.6)	73.6 (9.4)	−2.3(9.1)
1933–1942	65–74	73–82	665	75.2 (9.7)	73.4 (9.7)	−1.8(10.3)

BP at baseline in Tromsø Wave 6 (T0) and at follow-up in Tromsø Wave 7 (T1) and change. N = 4,465.

A decrease in cognition with age for both genders was observed as assessed by The Mini Mental State Examination (MMSE: test for dementia), the Digit Symbol Test of Wechsler Adult Intelligence Scale (WAIS for attention and speed) and the Twelve-word Test (for memory).

In 45–55 year-old-men, the performance of the Digit Symbol Test was poorer at the 8-year-follow up with increasing SBP and DBP. Such poorer performance was seen in the 65–74-old women on the Digit Symbol Test. However, higher cognition was seen with increased DBP and SBP in 45–55-year-old ladies.

Figure 4 below is illustrative of the trends in the cognition tests and the SBP (Systolic blood pressure) and DBP (Diastolic blood pressure) over eight years i.e. T1 (© 2020 Hestad, Engedal, Schirmer and Strand. *This is an open-access article distributed under the terms of the* ***Creative Commons Attribution License (CC BY)****. The use, distribution or reproduction in other forums is permitted, provided the original author(s) and the copyright owner(s) are credited and that the original publication in this journal is cited, in accordance with accepted academic practice. No use, distribution or reproduction is permitted which does not comply with these terms*):

The Digit Symbol Test

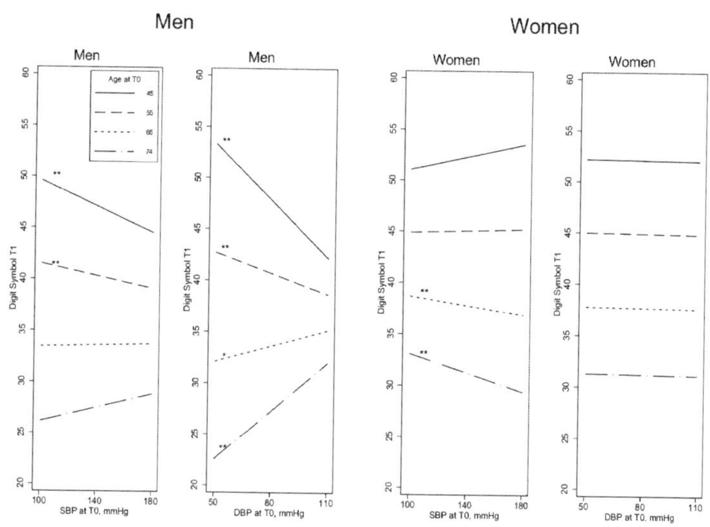

MMSE

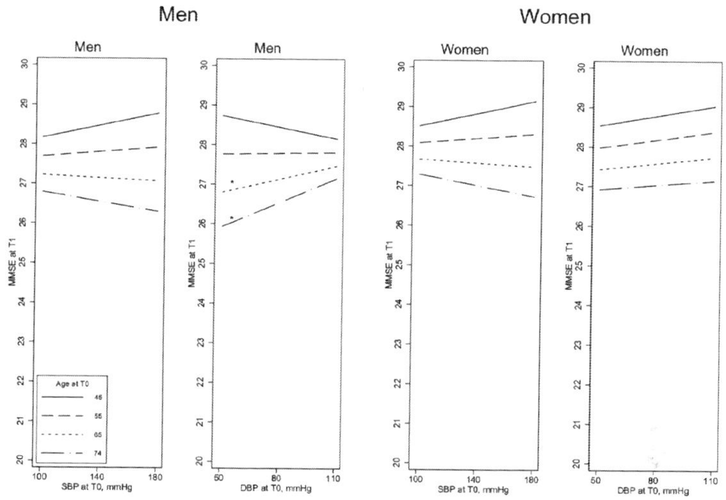

The Twelve-word Test

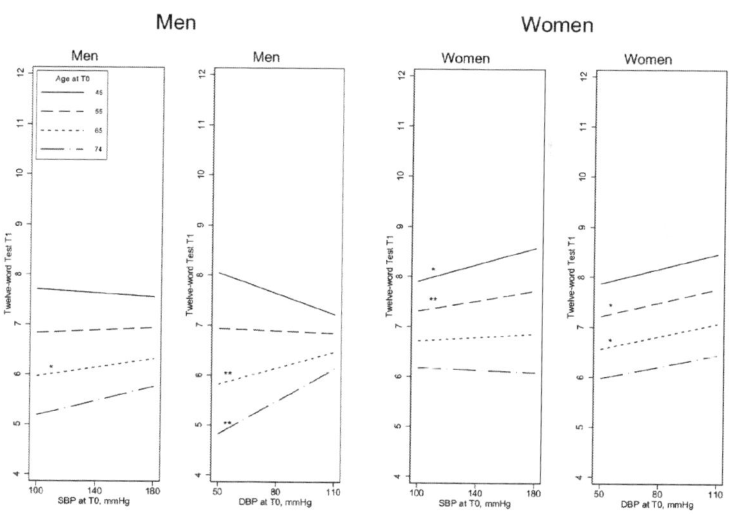

In a nutshell, the study described as "*the first time such a sex by BP interaction has been reported for cognition*" concluded that "*high BP at baseline for the men, but not the women, was clearly associated with reduced cognitive function 8 years later*" (Hestad et al, 2020).

A 2020-published article in *BMC Medical Genomics* by Sun and team looked at the use of Mendelian randomization (MR) to facilitate causal inference from observational data on the impact of high BP and cognition at mid-life. 1369 subjects of European ancestry from the multi-center cohort study called Coronary Artery Risk Development in Young Adults (CARDIA) were the participants. 25% of the participants (mean age=51 years at year 25) had hypertension with only 20% of them on medication.

The *TwoSampleMR* R package approach was used to look at the association with cognitive function and genetically predicted BP. It emerged that the Digit Symbol Substitution Test (DSST) score dipped by 0.49 to 0.67 for each mm of Hg rise in the BP. Further, the score on the Rey Auditory Verbal Learning Test (RAVLT) dipped by 0.7 and the Stroop Interference Test score was up by 2.3 for a 10 mmHg increment in genetically-predicted SBP that are all indicative of affected cognition.

Table 6 below illustrates the MR-based approach to explore the link between cognition at mid-life and blood pressure (Sun et al, 2020: Open Access:):

Key: IVW Inverse-variance weighted, *MR* Mendelian randomization, *SBP* Systolic Blood Pressure, *DBP* Diastolic Blood Pressure, *PP* Pulse Pressure, *DSST* Digit Symbol Substitution Test, *RAVLT* Rey Auditory Verbal Learning Test, *Stroop* The Stroop Interference test; [a] Represents the change in cognitive test score per 1-unit (mmHg) change in BP

	DSST			RAVLT			Stroop		
	SBP	DBP	PP	SBP	DBP	PP	SBP	DBP	PP
IVW									
Estimate[a]	−0.49	−0.66	−0.52	−0.04	−0.04	−0.07	0.15	0.18	0.16
95% CI	−0.79, −0.18	−1.20, −0.13	−0.99, −0.04	−0.11, 0.03	−0.16, 0.08	−0.18, 0.03	0.03, 0.32	−0.12, 0.48	−0.11, 0.43
***P*-value**	0.002	0.014	0.033	0.242	0.481	0.164	0.100	0.236	0.251
IVW after removing outliers									
Estimate[a]	−0.49	−0.77	−0.67	−0.07	−0.05	−0.07	0.23	0.24	0.27
95% CI	−0.80, −0.17	−1.03, −0.23	−1.16, −0.18	−0.14, 0.00	−0.16, 0.06	−0.17, 0.03	0.05, 0.41	−0.07, 0.54	−0.01, 0.56
***P*-value**	0.002	0.005	0.008	0.046	0.37	0.165	0.011	0.126	0.058

"That is the beauty and terror of midlife...one day...I began to wonder: Can I tweak my script?"

- Barbara Bradley Hagerty, Life Reimagined: The Science, Art, and Opportunity of Midlife

While the mechanisms that are associated with the change in cognition are under investigation, researchers Mansukhani, Kolla and Somers reviewed the plausible factors involved. Studies have pointed out the involvement of the Apolipoprotein ε4 (APOE4) allele. Other factors putatively indicated are inclusive of oxidative stress, dysfunction of the endothelium and hyperemia (excess blood supply). The remodeling of cerebral vessels that impacts the effective clearing of beta-amyloid is put forth as the predominant pathway. The metabolism of glucose has also been suggested to be lower in certain brain portions in hypertensive individuals.

Figure 5 below illustrates the pathways proposed for hypertension and cognitive impairment: here OSA refers to obstructive sleep apnea; RAAS, renin-angiotensin-aldosterone system; HPA, hypothalamic-pituitary axis (© 2019 Mansukhani, Kolla and Somers. *This is an open-access article distributed under the terms of the* ***Creative Commons Attribution License (CC BY)****. The use, distribution or reproduction in other forums is permitted, provided the original author(s) and the copyright owner(s) are credited and that the original publication in this journal is cited, in accordance with accepted academic practice. No use, distribution or reproduction is permitted which does not comply with these terms*):

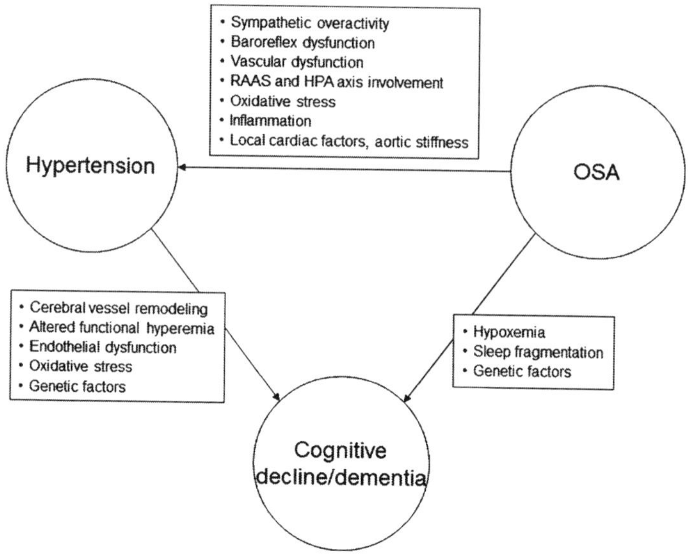

4.8 THE BLUES: DEPRESSION

White and team (2006) scrutinized the level of depression, anxiety, quality of life and cognition in 46 pulmonary hypertension patients from the pulmonary hypertension clinic at LDS Hospital, in Salt Lake City, Utah and published their results in *Respiratory Research*. While the Beck Depression Inventory (BDI) and Beck Anxiety Inventory (BAI) were utilized to assess depression and anxiety, respectively, the quality of life of the pulmonary arterial hypertension (PAH) patients was evaluated by the Medical Outcome Study 36-Item Short Form Health Survey (SF-36). Various aspects of cognition were scrutinized by the National American Adult Reading Test (NAART: intellect), the Logical Memory I and II subtests from the Weschler Memory Scale-III (WMS-III), the Trail Making Test Parts A and B, Wisconsin Card Sorting Test, and Hayling Sentence Completion Test (for functioning), the Stroop Test, Ruff 2&7 Selective Attention Test and Digit Span from the WMS-III (for attention), Symbol Digit subtest (processing speed) and the Grooved Pegboard and Finger Tapping tests (for motor function).

Moderate to severe symptoms of depression were seen in 26% of the patients while severe anxiety was seen in 19.6% of PAH patients. The overall quality of life was also lower in the PAH patients. There was an impact noted on cognition with affected motor function in 56.5%; compromised memory in 41.3%; slower mental processing in 17.4%; dysfunction in 15.2% and lower attention in 13.5% of the patients.

Table 7 below illustrates the impact of PAH on ***TOP***: depression and quality of life and cognition (White et all, 2006: Open Access: *This article is published under license to BioMed Central Ltd. This is an Open Access article is distributed under the terms of the Creative Commons Attribution License (https://creativecommons.org/licenses/by/2.0*), which permits unrestricted use, distribution, and reproduction in any medium, provided the original work is properly cited):

	Total PAH group (N = 46) Mean ± SD	**Normals**	**Cognitive Sequelae (N = 27) M ± SD**	**No Cognitive Sequelae (N = 19) M± SD**	**Cognitive Sequelae vs. No Cognitive Sequelae P value**
SF-36 Scores					
General Health	37.6 ± 19.2	77	37.7 ± 18.6	37.6 ± 20.7	0.99
Physical Health	36.6 ± 23.5	82	35.7 ± 18.3	37.8 ± 30.3	0.78
Role Physical	37.2 ± 37.9	81	28.6 ± 35.2	50.0 ± 39.3	0.06
Role Emotional	69.6 ± 42.5	83	61.7 ± 46.0	81.5 ± 34.7	0.13
Social Functioning	55.1 ± 29.4	82	52.5 ± 25.6	58.7 ± 34.6	0.52
Bodily Pain	61.2 ± 24.3	79	60.8 ± 26.0	61.8 ± 22.3	0.89
Vitality	36.7 ± 22.1	65	35.7 ± 22.0	38.1 ± 23.0	0.74
Mental Health	52.0 ± 35.5	80	55.0 ± 33.9	47.6 ± 38.5	0.52

BOTTOM: the effects of PAH on cognition (White et all, 2006: Open Access: *This article is published under license to BioMed Central Ltd. This is an Open Access article is distributed under the terms of the Creative*

Key: IQ = Intelligence Quotient; NAART = North American Adult Reading Test; and WCST = Wisconsin Card Sorting Test

	Total Group M ± SD	**Cognitive Sequelae M ± SD**	**No Cognitive Sequelae M ± SD**	**Cog. Sequelae vs. No Cog. Sequelae P value**
WASI				
Full Scale IQ	103 ± 12.0	100 ± 13.5	107 ± 7.8	0.06
NAART Estimated Premorbid IQ				
Full Scale Intelligence Quotient	111 ± 9.1	109 ± 9.9	114 ± 6.9	0.13
Memory				
Rey Auditory Verbal Learning Test (Raw Scores)				
Trial 1	6.2 ± 1.9	5.7 ± 1.8	6.8 ± 1.7	**0.04**
Trial 5	12.3 ± 2.3	11.5 ± 2.5	13.3 ± 1.4	**0.01**
Immediate Recall	10.5 ± 3.2	9.7 ± 3.3	11.5 ± 2.8	0.07
Delayed Recall	10.3 ± 3.3	9.8 ± 3.6	10.8 ± 2.7	0.29
Logical Memory (Percentile)				
Immediate Recall	41.4 ± 27.8	32.5 ± 26.1	54.0 ± 25.6	0.08
Delayed Recall	41.9 ± 27.4	32.2 ± 24.4	55.8 ± 25.8	**0.03**
Rey Complex Figure Test (T-scores)				
Immediate Recall	49.8 ± 14.9	46.8 ± 15.1	54.1 ± 13.8	0.88

Delayed Recall	47.9 ± 14.7	45.4 ± 16.0	51.4 ± 12.0	0.13
Attention				
Letter-Number Sequencing (Scaled Score)	9.5 ± 2.8	8.9 ± 2.9	10.2 ± 2.4	0.27
Digit Span (T-score)	59.9 ± 10.4	58.6 ± 10.8	61.7 ± 9.7	0.32
Ruff 2&7 Test (T-scores)				
Total Speed	58.0 ± 10.9	56 ± 11.6	61.0 ± 9.3	0.11
Total Accuracy	44.8 ± 12.5	45 ± 14.7	44.5 ± 8.6	0.89
Stroop Color-Word T-score	48.2 ± 14.5	47 ± 14.5	49.8 ± 14.7	0.91
Executive Function				
Trail Making Test (T-scores)				
Part A	50.3 ± 10.1	49.3 ± 10.3	51.6 ± 9.8	0.45
Part B	49.5 ± 7.9	49.5 ± 7.7	49.5 ± 8.3	0.98
WCST				
Total Score (T-score)	46.1 ± 9.9	44.8 ± 8.2	48.1 ± 12	0.28
Perseverative Errors (T-score)	48.4 ± 7.9	46.4 ± 8.4	51.4 ± 6.0	**0.04**
Number Categories Completed	5.2 ± 1.3	4.9 ± 1.4	5.6 ± 0.8	0.07
Hayling Sentence Completion Test				
Total Score (Scaled Score)	5.6 ± 1.7	5.5 ± 2.0	5.8 ± 0.9	0.55
Mental Processing Speed				
Symbol Digit (Raw Score)	52.6 ± 13.8	51.7 ± 14.8	50.8 ± 17.5	0.86

Motor Function (T-scores)				
Finger Tapping Right hand	58.1 ± 12.1	55.4 ± 12.9	62.6 ± 9.4	**0.05**
Finger Tapping Left hand	61.7 ± 9.3	59.4 ± 8.2	65.9 ± 9.9	**0.03**
Grooved Peg Board Right hand	43.7 ± 11.5	37.8 ± 10.4	53.2 ± 5.0	**<0.001**
Grooved Peg Board Left hand	42.8 ± 12.7	41.0 ± 13.8	46.1 ± 9.7	0.21

Research published in 2009 explored the link between anxiety/depression and hypertension on the basis of the South African Stress and Health (SASH) study: part of the World Health Organization World Mental Health Survey Initiative: a cross-sectional survey of mental health in South African adults. The sampling was between 2002 and 2004 with a self-reported diagnosis of hypertension and 12-month diagnosing of depression employing the Diagnostic and Statistical Manual of Mental Disorders 4th edition (*DSM-IV*).

Reporting of hypertension was more likely in women than men by a factor of 2.6. Individuals with another chronic condition were 6.7 times more likely to report a lifetime diagnosis of hypertension vs. participants without a hypertension diagnosis (p-<0.001). Lower education and unemployment were more influencing factors for hypertension diagnosis.

Analyses revealed that of the 4351 adults, a 12-month anxiety disorder was seen in 8%; 12-month depressive disorder in 5% a 12-month and 12-month comorbid anxiety-depression in 1% of the participants. 43%of the participants reported a chronic physical condition and 79% of the participants with hypertension had at least one chronic condition.

Table 8 below is illustrative of the observations of ***TOP***: Hypertension diagnosis-based sample stratification (© 2009 Grimsrud et al. *This is an open-access article distributed under the terms of the Creative Commons Attribution License, which permits unrestricted use, distribution, and reproduction in any medium, provided the original author and source are credited*):

	Total	Hypertension diagnosis	No hypertension diagnosis	p-value
	%	%	%	
N (%)	4351	767 (16.7)	3584 (83.3)	
Demographic Characteristics				
Mean Age	37.0	50.3	34.3	-
Age categories: 18–29	39.1	9.0	45.1	<0.001
30–39	22.1	13.5	23.8	
40–49	18.1	24.2	16.9	
50+	20.7	53.3	14.2	
Sex: Male	46.3	28.0	50.0	<0.001
Female	53.7	72.0	50.0	
Race: Black	76.2	76.1	76.3	0.271
Coloured	10.4	11.7	10.2	
White	10.0	10.2	9.9	
Indian/Asian	3.4	2.0	3.6	
Currently married	50.1	54.0	47.1	<0.001
Location: Rural	38.4	41.6	37.7	0.074
Urban	61.6	58.5	62.3	
Socioeconomic status				
Education: None	6.8	12.7	5.6	<0.001
Grade 1–7	19.1	33.9	16.2	
Grade 8–11	35.4	32.9	35.9	
Completed high school	23.5	11.4	25.9	
Post-high school education	15.3	9.2	16.5	
Employed	31.0	26.0	32.0	0.020
Household income (mean)	59403.9			
R0	13.7	13.3	13.7	0.244
R1–5000	29.5	32.9	28.9	
R5001–25000	15.4	16.7	15.1	
R25000–100000	19.6	17.4	20.1	
R100001+	21.8	19.7	22.3	
Assets owned by household: 0–5	39.3	38.3	39.5	
6–12	37.4	41.0	36.7	
13–17	23.3	20.7	23.8	0.177
SES Quartile: 0	22.5	36.0	19.8	<0.001
1	24.5	22.1	24.9	
2	25.0	20.8	25.9	
3	28.0	21.0	29.4	
Other				
Alcohol: Lifetime – None	77.8	79.1	77.5	0.157
Rare	5.8	7.0	5.5	
Moderate	7.5	5.7	7.9	
Heavy	8.9	8.2	9.1	
Alcohol: Current – None	65.7	71.8	64.5	0.053
Rare	10.2	8.7	10.5	
Moderate	16.8	13.5	17.4	
Heavy	7.3	6.0	7.6	
Smoking				
Lifetime smoker	30.0	26.1	30.8	0.033
Current smoker	23.8	15.9	25.4	<0.001
Chronic physical conditions	42.6	78.5	35.4	<0.001
Vascular	11.2	40.2	5.4	<0.001
Other	39.4	69.9	33.3	<0.001
Traumatic life events				
None	26.6	17.3	28.5	<0.001
1–2	49.9	52.5	49.4	
3–4	10.8	11.9	10.6	
5 or more	12.7	18.3	11.5	

doi:10.1371/journal.pone.0005552.t001

BOTTOM: 12-month depression and anxiety trends in the sample (© 2009 Grimsrud et al. *This is an open-access article distributed under the terms of the Creative Commons Attribution License, which permits unrestricted use, distribution, and reproduction in any medium, provided the original author and source are credited*):

	No depression or anxiety	Anxiety	OR (95% CI)	Major Depression	OR (95% CI)	Comorbid anxiety-depression	OR (95% CI)
	%	%		%		%	
Prevalence	87.8	8.1		4.9		1.4	
Mean Age	36.8	38.4		37.6		39.5	
Age categories:18–29	39.8	33.3	1.0	32.2	1.0	22.9	1.0
30–39	21.9	23.2	1.3 (1.0–1.7)	24.4	1.4 (0.9–2.1)	30.8	2.4 (1.0–6.0)
40–49	17.6	21.2	1.4 (1.0–2.1)	23.3	1.6 (1.1–2.5)	20.9	2.1 (1.0–4.2)
50+	20.7	22.3	1.3 (0.8–2.0)	20.2	1.2 (0.8–1.8)	25.4	2.1(1.0–4.5)
Sex: Male	48.4	30.1	1.0	29.2	1.0	11.3	1.0
Female	51.7	69.9	2.2 (1.6–2.9)	70.8	2.3 (1.6–3.2)	88.7	7.4 (3.4–16.0)
Race: Black	76.2	79.9	1.0	71.8	1.0	79.8	1.0
Coloured	10.1	11.5	1.1 (0.7–1.7)	13.4	1.4 (0.9–2.1)	9.7	0.9 (0.3–2.4)
White	10.4	6.5	0.6 (0.2–1.6)	7.6	0.8 (0.3–2.1)	7.7	0.7 (0.1–5.9)
Indian/Asian	3.2	2.1	0.6 (0.3–1.3)	7.2	2.4 (1.1–5.0)	2.8	0.8 (0.2–3.1)
Currently married	50.1	48.6	0.9 (0.7–1.2)	53.1	1.1 (0.9–1.5)	50.9	1.0 (0.6–1.7)
Location: Urban	61.2	66.1	1.2 (0.9–1.6)	65.2	1.2 (0.9–1.7)	76.2	2.0 (1.1–2.6)
Hypertension diagnosis	15.7	26.9	2.0 (1.4–2.8)	21.2	1.4 (1.0–2.0)	31.6	2.4 (1.2–5.1)
Lifetime smoker	29.0	34.6	1.3 (1.0–1.6)	38.9	1.6 (1.1–2.2)	27.0	0.9 (0.5–1.7)
Current smoker	23.2	24.6	1.1 (0.9–1.4)	31.1	1.5 (1.0–2.2)	19.0	0.8 (0.4–1.6)
Other chronic physical condition	40.3	59.8	2.2 (1.7–2.8)	59.8	2.2 (1.6–3.1)	67.8	3.1 (1.6–6.2)
Vascular	10.2	20.0	2.2 (1.5–3.3)	16.3	1.7 (1.1–2.6)	21.8	2.5 (1.0–5.8)
Other	37.1	57.8	2.3 (1.8–3.0)	56.2	2.2 (1.5–3.1)	64.6	3.1 (1.6–6.1)
Traumatic life events: None	28.3	16.8	1.0	10.0	1.0	10.2	1.0
1–2	50.2	45.0	1.5 (1.0–2.2)	51.5	2.9 (1.7–4.9)	40.6	2.3 (0.9–5.6)
3–4	10.4	13.0	2.1 (1.3–3.5)	15.7	4.2 (2.4–7.6)	19.9	5.3 (1.9–14.7)
5 or more	11.1	25.2	3.8 (2.6–5.7)	22.8	5.8 (3.6–9.4)	29.4	7.4 (2.7–20.3)

doi:10.1371/journal.pone.0005552.t002

Participants with agoraphobia, generalized anxiety disorder and panic disorder showed a ubiquitous occurrence of hypertension: the likelihood of being diagnosed with hypertension was 1.68 times more in individuals with a 12-month mental disorder as opposed to those without a 12-month mental disorder. Finally, 12-month anxiety disorders were associated with hypertension and other chronic conditions. In a nutshell, this work described as the "*first population-based study from sub-Saharan Africa to document an association between anxiety and depressive disorders and hypertension*" reported the association of anxiety/depression with hypertension and chronic health issues.

A cross-sectional community-based study at Monastir (Tunisia) explored the extent of depression in the elderly employing a mini-Geriatric Depression Scale mini-GDS in 598 subjects (mean age=72.3 ± 7.4 years). The prevalence of hypertension was 52%: 55.5% in women and 45% in men. 81% of the participants were aware of the condition with 78.4% of them taking medication. Such predominance of arterial hypertension in elderly people warrants awareness not only at the individual level but also at the community level along with the formulation of appropriate approaches to tackle the condition by the healthcare system.

Table 9 below shows the trend in arterial hypertension along with associated factors across a community of elderly people in Tunisia (Hammami et al, 2011: *This article is published under license to BioMed Central Ltd. This is an Open Access article distributed under the terms of the Creative Commons Attribution License (http://creativecommons.org/licenses/by/2.0), which permits unrestricted use, distribution, and reproduction in any medium, provided the original work is properly cited*):

Key: [a] Figures in parentheses are percentages [b] chi2 test [c] not statistically significant

Variable	**N**	**Prevalence of Hypertension**	**p value** [b]
Gender			
Female	396	220 (55.5) [a]	
Male	202	91 (45)	P < 0.05
Age (years)			
< 70	227	117 (51.5)	
70-74	142	78 (55)	
75-79	122	67 (55)	
≥ 80	107	49 (45.8)	P < 0.01
Schooling			
Illiterate	460	248 (54)	
Kuttab (koranic schools)	72	37 (51.3)	
Primary	34	16 (47)	
Secondary or more	32	10 (31.2)	p = 0.08
Dependency			
Intense	57	39 (68.4)	
Moderate	150	86 (57.3)	
Autonomy +	391	186 (47.5)	P < 0.01
Place of residence			
Urban	516	270 (52.3)	NS [c]

Rural	82	41 (50)	
Depression			
Mini-GDS < 1	462	232 (50.2)	
Mini-GDS ≥ 1	136	79 (58)	NS
Marital status			
Married	359	187 (52)	
Single	239	124 (51.8)	NS
Physical activity			
Yes	23	8 (34.7)	
No	575	303 (52.6)	NS
Diabetes			
Absent	434	195 (45)	
Present	164	116 (70.7)	P< 0.001
Overweight (kg/m ^{2}) (n = 531)			
BMI< 25	106	31 (29.2)	
BMI: 25-29.9	165	81 (49)	
BMI ≥ 30	260	167 (64.2)	p = 0.000
Abdominal circum-ference			
(n = 531)			
Normal	142	47 (33)	
Increased	389	235 (60.4)	P = 0.000

In a 2013-published report, the extent of depression was evaluated in 158 patients [138 with pulmonary Hypertension (PH) and 20 with inoperable chronic thromboembolic PH (CTEPH)]: 113 females and 45 males (mean age = 56± 16 years). The assessment was based on the Patient Health Questionnaire (PHQ-9) and Generalized Health Anxiety Disorder 7-item questionnaire (GAD-7). The researchers concluded that a significant impairment of life quality was observed with depression and anxiety in PH patients with the prevalence of depression or anxiety in 26.5% of females and 13.3% of males. Overall, the team stated that "*MD (mental disorders) in PH appears to be not only underdiagnosed but also undertreated*".

Table 10 below depicts the results of the grim scenario of mental health in PH patients (Harzheim et al, 2013: *Open Access: This article is published under license to BioMed Central Ltd. This is an Open Access article is distributed under the terms of the Creative Commons Attribution License (https://creativecommons.org/licenses/by/2.0*), which permits unrestricted use, distribution, and reproduction in any medium, provided the original work is properly cited):

Symptoms	**Anxiety disorder (GAD-7)**	**Depression disorder (PHQ-9)**
No symptoms	11 (7.0%)	9 (5.7%)
Minimal symptoms	75 (47.5%)	62 (39.2%)
Mild symptoms	55 (34.8%)	55 (34.8%)
Moderate symptoms	13 (8.2%)	21 (13.3%)
Serious/severe symptoms	4 (2.5%)	11 (7.0%)
Daily routine difficulty		
	No or mild MD	Anxiety or depression
		disorder (moderate to severe)
Daily routine difficulty GAD-7 (Anxiety)		
n	105	32
Not difficult at all	41 (39.0%)	1 (3.2%)
Somewhat difficult	54 (51.4%)	17 (54.8%)
Very difficult	9 (8.6%)	8 (25.8%)
Extremely difficult	1 (1.0%)	5 (16.1%)
Daily routine difficulty PHQ-9 (Depression)		
n	103	33
Not difficult at all	29 (28.2%)	0 (0.0%)
Somewhat difficult	62 (60.2%)	14 (42.4%)
Very difficult	11 (10.7%)	13 (39.4%)
Extremely difficult	1 (1.0%)	6 (18.2%)

A cross-sectional study published in the *International Journal of Hypertension* in an outpatient clinic in Andkhoy, Afghanistan between December 2015 and August 2016 assayed the factors linked to distress in hypertensive patients (more than 50% over 60 years of age). Of the 234 participants, 58.1% had depression and 42.3% had anxiety as measured by the Hospital Anxiety and Depression Scale (HADS) questionnaire. Further analyses of associated factors are presented below in table 11 (Mohammad Shoaib Hamrah et al, 2018: Copyright © 2018 Mohammad Shoaib Hamrah et al. *This is an open access article distributed under the Creative Commons Attribution License, which permits unrestricted use, distribution, and reproduction in any medium, provided the original work is properly cited*):

The Link with Clinical and Socio-economic Factors

	Depression		**Anxiety**		**Comorbid depression-anxiety**	
	n=136	P *value*	n=99	P *value*	n=66	P *value*
Age, mean, years mean (SD)	63.8±15.0	0.915	54.6±12.7	0.019	55.7±15.3	0.04
Gender		0.099		<0.001		0.573
Male,%	83 (61.0%)		79 (79.8%)		45 (68.2%)	
Female, %	53 (39.0%)		20 (20.2%)		21 (31.8%)	
Marital status		0.395		0.003		0.005
Married	119 (87.5%)		76 (76.8%)		49 (74.2%)	
Single	11 (8.1%)		17 (17.2%)		13 (19.6%)	
Others	6 (4.4%)		6 (6.0%)		4 (6.1%)	
Age (in years)		<0.001		0.001		0.019
≤39	12 (8.8%)		18 (18.2%)		13 (19.6%)	
40-69	36 (26.5%)		36 (36.4%)		23 (34.8%)	
>60	88 (64.7%)		45 (45.4%)		30 (45.4%)	
Level of education		0.193		0.414		0.026
Illiterate	85 (62.5%)		58 (58.6%)		33 (50.0%)	

Primary/private education	8 (5.9%)		6 (6.0%)		5 (7.6%)	
Secondary	27 (19.8%)		17 (17.2%)		18 (27.3%)	
High school/ higher	16 (11.8%)		18 (18.2%)		10 (15.1%)	
Occupation		0.086		<0.001		0.075
Employed	30 (22.1%)		12 (12.1%)		9 (13.6%)	
Unemployed	54 (39.7%)		53 (53.5%)		33 (50.0%)	
Housewife	27 (19.8%)		25 (25.2%)		15 (22.7%)	
Others	25 (18.4%)		9 (9.1%)		9 (13.6%)	
Body mass index, mean (SD)	24.5±4.4	0.838	23.3±4.24	0.016	23.7±4.4	0.545
Diabetes mel-litus	61 (44.8%)	<0.001	40 (40.4%)	0.006	37 (56.1%)	<0.001
Smoking		0.128		<0.001		0.055
Yes	49 (36.0%)		53 (53.5%)		33 (50.0%)	
No	87 (64.0%)		46 (46.5%)		33 (50.0%)	
Left ventricular hypertrophy	80 (58.8%)	0.236	45 (45.4%)	0.008	28 (42.4%)	0.011
Heart failure	22 (16.2%)	0.070	11 (11.1%)	0.503	8 (12.1%)	0.841
Stroke	11 (8.1%)	0.016	3 (3.0%)	0.213	3 (4.5%)	0.800
Myocardial infarction	34 (25.0%)	0.004	13 (13.1%)	0.057	12 (18.2%)	0.879

What Influenced Anxiety in Hypertensive Patients

Key: CI, confidence interval; OR, odds ratio; DM, diabetes mellitus; MI, myocardial infarction.

Variables	Anxiety n (%)	OR	(95% CI)	P *value*
Age ≤ 39	18 (18.2)	1.0		
40-60	36 (36.4)	2.61	(0.866-7.88)	0.088
>60	45 (45.4)	5.0	(1.66-15.17)	0.004
Occupation, Employed	12 (12.1)	1.0		
Housewife	25 (25.2)	1.96	(0.62-6.24)	0.253
Others	9 (9.1)	0.287	(0.106-0.77)	0.014
Unemployed	53 (53.5)	0.431	(0.148-1.255)	0.123
Gender, Male	20 (20.2)	1.0		
Female	79 (79.8)	4.25	(2.07-8.73)	<0.001
Smoking, No	46 (46.5)	1.0		
Yes	53 (53.5)	2.95	(1.55-5.63)	0.001
Comorbidities				
Only hypertension	1.0			
with DM	40 (40.4)	4.90	(1.19-20.22)	0.028
DM with MI or Stroke	9 (9.1)	5.66	(1.37-23.42)	0.017

The Risk of Depression and Its Factors

	Depression			**Comorbid depression-anxiety**		
Variables	**OR**	**95% CI**	**P** *value*	**OR**	**95% CI**	**P** *value*
Age ≤ 39	1.0					
40-60	3.99	(1.41-11.28)	0.009	2.63	(1.00-6.94)	0.05

>60	4.04	(2.07-7.89)	<0.001	3.50	(1.37-8.97)	0.009
Comorbidities						
only hypertension	1.0					
with DM	22.74	(2.69-192.36)	0.004	10.24	(2.79-37.67)	<0.001
DM and MI or stroke	2.91	(0.34-25.03)	0.331	3.96	(1.10-14.25)	0.035

(Mohammad Shoaib Hamrah et al, 2018)

A cross-sectional study done during November 2017 to February 2018 at the rural health center Marapaddi, Kanayakumari, India explored the prevalence of depression in 200 patients was published in the *National Journal of Community Medicine.* The assessment of depression was on the basis of the Patient Health Questionnaire – 9 (PHQ-9) that revealed that 41% of hypertensive participants (mean age=66.16±8.62 years) had depression. Further analyses revealed the involvement of factors such as comorbidities, lack of medication, family history, and socio-economic status as displayed below in table 12 (Ashok and Ghosh, 2019: *N You can Share — copy and redistribute the material in any medium or format; Adapt — remix, transform, and build upon the material for any purpose, even commercially*):

Table 1: Prevalence of depression among the study subjects, N=200.

Depression	Participants (n) (%)
No depression	118 (59)
Mild depression	57 (28.5)
Moderate depression	14 (7)
Moderately severe depression	6 (3)
Severe depression	5 (2.5)
Total	200 (100)

Table 2: Factors found significantly associated with depression among the study subject

Factors	Depressive (n=82) (%)	Non depressive (n=118) (%)	p value	OR (CI)
Age				
< 65 years	20 (29)	49 (71)	0.012	0.454 (0.244-0.847)
>65 years	62 (47.3)	69 (52.7)		
Gender				
Females	61 (46.9)	69 (53.1)	0.020	2.063 (1.114 -3.821)
Males	21 (30.0)	49 (70.0)		
Marital status				
Living with spouse	37 (32.7)	76 (67.3)	0.002	0.459 (0.256-0.805)
Single	45 (37.7)	42 (48.3)		
SES				
BPL	61 (50.4)	60(49.6)	<0.001	2.808 (1.521 - 5.185)
APL	21(26.6)	58 (73.4)		
Family history				
Present	61 (52.5)	53(46.5)	<0.001	3.562 (1.972 - 6.585)
Absent	21 (24.4)	65 (75.6)		
BP control				
Uncontrolled	53 (60.2)	35 (39.8)	<0.001	4.334 (2.377 - 7.904)
Controlled	29 (25.9)	83 (74.1)		
Duration of BP				
< 6 years	14 (11.3)	110 (88.7)	<0.001	0.015 (0.006-0.038)
>6 years	68 (89.5)	8 (10.5)		
On Medication				
No	37 (61.7)	23 (38.3)	<0.001	3.396 (1.809-6.375)
Yes	45 (32.1)	95 (67.9)		
Co-morbidities				
Present	67 (48.2)	72 (51.8)	0.002	2.854 (1.454-5.583)
Absent	15 (24.6)	46 (75.4)		

Among the complications associated with pregnancy, hypertensive disorders of pregnancy (HDP) account for 10% of pregnancies. A review by Roberts, Davis and Homer in *Frontiers in Cardiovascular Medicine* presented studies that explored studies on depression and HDP as presented below in table 13 ():

References	Study design recruitment	Location	Instrument	When measured	Cohort (N)	Results		
						Depression	Anxiety	PTSD

Abedian et al. (40)	Descriptive study	Iran	BDI STAI PPQ	2 weeks pp 6 weeks pp	PE (n = 100) 56 primiparous 44 multiparous	No significant difference at both time points (no figures given)	No significant difference at both time points (no figures given)	Mean scores: Primip 4.4 Multip 5.4 (p = 0.05) Prevalence: Primip 21% Multip 32% (p = 0.24) (unclear which time these results are from)
Baecke et al. (28)	Cohort study Recruited after birth	Netherlands 1 Hospital	BDI IES >25 STAI	6–18 months pp	Preterm PE (n = 48) Term PE (n = 18) PTB (n = 32) Term birth (n = 72)	Mean scores: Preterm PE 2.7 Term PE 1.3 PTB 1.2 Term birth 1.8 (P = 0.11)	Mean scores: Preterm PE 39.9 Term PE 39.6 PTB 35.4 Term birth 37.2 (P = 0.27)	Prevalence: Preterm PE 44% Term PE 11% PTB 41% Term birth 11% (P < 0.001) Mean scores: Preterm PE 25.3 Term PE 10.5 PTB 20.0 Term birth 12.7 (P < 0.001)
Blom, et al. (33)	Prospective cohort study Recruited antenatally	Netherlands Women in Rotterdam with EDC 04/2002-01/2006	EPDS > 12	2 months pp	PE (n = 183) Other (n = 3917) Normal (n = 770)	Prevalence: PE 16.9% GH 7.6% PE associated with increased risk of PND (OR 2.58, 95% CI 1.30–5.14)		

Brusse et al. (34)	Cohort study Recruited after birth	Netherlands 1 hospital	CES-D STAI	3–7 months pp	PE (n = 10) Normal (n = 10)	Median scores: PE 7.5 Normal 8 (P = 0.67)	Median scores: PE 39 Normal 31 (P = 0.21)	
Chen et al. (41)	Retrospective	China 1 hospital	EPDS ≥ 10	6 weeks pp	PE (n = 90) N (n = 90)	Mean scores: PE 7.27 N 4.42 (p = 0.05) Prevalence: PE 27% N 12% (p = 0.014)		
Engelhard et al. (29)	Cohort study Recruited after birth	Netherlands 1 Hospital	BDI>15 PSS	Within 2 years of hospital admission	Preterm PE (n = 18) PTB (n = 29) Term PE (n = 23) Term birth (n = 43)	Prevalence: Preterm PE 33% PTB 24% Term PE 26% Term birth 7%		Prevalence: Preterm PE 28% PTB 28% Term PE 17% Term birth 0%
Fields et al. (42)	Cohort study Women identified via medical record linkage	US	BDI BAI	Once, average 34.7 years since index pregnancy	PE (n = 40) N (n = 40)	Mean scores: PE 4 N 2 (p = 0.64)	Mean scores: PE 3 N 105 (p = 0.58)	

Gaugler-Senden et al. (35)	Cohort study Recruited after birth	Netherlands 1 hospital birthed 1993–2004	ZDS IES >19	In 2008 asked to recall symptoms in pp period and at current time	PE (n = 104) Normal (n = 78)	Mean scores: Postpartum recall: PE 11.78 N 11.37 (P = 0.59) In 2008: PE 5.4 Normal 5.51 (P = 0.87)		Mean scores: Postpartum recall: PE 29.72 Normal 27.53 (P = 0.41) Currently: PE 28.66 Normal 25.69 (P = 0.02)
Habli et al. (43)	Cohort study	US Advertised on HELLP Syndrome Society website	Study designed survey	1–31 years after birth	HELLP (n = 128)	Prevalence: 32%	Prevalence: 26%	
Hoedjes et al. (44)	Prospective cohort study Recruited within 6 weeks pp	Netherlands 4 hospitals	EPDS ≥ 10 EPDS ≥ 13	6 weeks pp 12 weeks pp 26 weeks pp	Mild PE (n = 39) Severe PE (n = 122)	Prevalence: (≥10) at any time: Mild PE 23.1% Severe PE 44.3% (P = 0.018) (≥13) at any time: Mild PE 15.4% Severe PE 26.2% (P = 0.165)		

Hoedjes et al. (30)	Prospective cohort study Recruited within 6 weeks pp	Netherlands 4 hospitals	Self-rating Inventory for PTSD	6 weeks pp 12 weeks pp	Mild PE (n = 35) Severe PE (n = 114)			Prevalence total study population: 6 weeks pp 8.6% 12 weeks pp 5.1% (P = 0.083)
Mautner et al. (45)	Prospective cohort study Recruited antenatally between 24 and 37 weeks gestation	Germany 1 tertiary referral hospital	EPDS ≥10	24–37 weeks gestation 2–5 days pp 3–4 months pp	HDP (n = 18) PTB (n = 32) Normal (n = 29)	Mean scores: 2–5 days pp HT 7.83 PTB 9.91 Normal 4.69 3–4 months HT 3.67		
			PTB 6.53 Normal 5.48			Prevalence in HT: 39% antenatally 27% 2–5 days pp 17% 3–4 months pp		
Mommersteeg et al. (46)	Cohort study	Netherlands 1 hospital	PHQ-9 ≥5 GAD-7 ≥5	Once, average 14 years after index pregnancy	PE (n = 265) N (n = 268)	Prevalence: PE 27% N 23% (p = 0.27) Mean scores: PE 3.6 N 2.9 (p = 0.03)	Prevalence: PE 31% N 27% (p = 0.33) Mean scores: PE 3.6 N 3.5 (p = 0.72)	

Porcel et al. (47)	Cross sectional online survey First pregnancy 1990–2010	World-wide PE Foundation US based	BSSS for DSM IV ≥4	Once	HDP (n = 1,076) Normal (n = 372)			Prevalence: HDP 43% Normal 14% (P ≤ 0.01)
Postma et al. (48)	Observational cohort study Recruited 1–27 years after index pregnancy	Netherlands	HADS	Once, at least 12 months after the index pregnancy	Eclampsia (n = 46) PE (n = 51) Normal (n = 48)	Mean scores: Eclampsia 5 PE 4 Normal 3 (P = 0.02)	Mean scores: Eclampsia 7 PE 6 Normal 5 (P < 0.005)	
Postma et al. (49)	Retrospective cohort study	Netherlands	HADS	1–27 years since index pregnancy (mean = 6 years)	PE (n = 90) Normal (n = 47)	Mean total scores: PE 11 N 8 (p < 0.001)	Mean total scores: PE 11 N 8 (p < 0.001)	
Stramrood et al. (31)	Prospective cohort study Recruited antenatally when hospitalized (PE) or at 38 weeks gestation (controls)	Netherlands 1 hospital 1 midwifery practice	BDI > 20 PSSR-SR	Pregnancy 6 weeks pp 15 months pp	PE (n = 63) Normal (n = 65)	Prevalence: Pregnancy PE 19% Normal 7.7% 6 weeks pp PE 10.5% Normal 6.2% 15 months pp PE 6.8% Normal 0%		Prevalence: 6 weeks pp PE 10.5% Normal 3% 15 months pp PE 11.4% Normal 0%

BDI, Beck Depression Inventory; BSSS, Breslau Short Screening Scale; CES-D, Center for Epidemiological Studies Depression Scale; CI, Confidence Interval; DSM IV, Diagnostic Statistical Manual of Mental Disorders 4th Edition; EDC, expected date of confinement; EPDS, Edinburgh Postnatal Depression Scale; GH' gestational hypertension; HADS, Hospital Anxiety and Depression Scale; HELLP, Hemolysis Elevated Liver enzymes Low Platelets; HDP, hypertensive disorder of pregnancy; IES, Impact of Event Scale; OR, Odds Ratio; PE, preeclampsia; PND, postnatal depression; pp, postpartum; PSS, Post-traumatic stress Symptom Scale; PSSR-SR, PTSD Symptom Scale self-report questionnaire; PTB, preterm birth; PTSD, post-traumatic stress disorder; SCL-90, 90 item Symptom Check List; STAI, State Trait Anxiety Inventory; ZDS, Zung Depression Scale.

Overall, an increasing trend of depression, anxiety, and PTSD (post-traumatic stress disorder) after hypertensive disorders of pregnancy (HDP) was reported by the researchers. This warrants deeper research to help awareness and prompt screening of such conditions.

Looking at the prevalence of depression with hypertension, what is the way out? Below are few approaches as ***representative solutions*** as outlined by research (***this is not meant to be diagnostic***):

A 2015-published article in *Evidence-Based Complementary and Alternative Medicine* reported the effects of **progressive muscle relaxation (PMR)** on depression and quality of life in pulmonary arterial hypertension (PAH) patients. This randomized controlled study was inclusive of a PMR group and a control group (65 participants in each: final analyses of 55 cases in the PMR group and 59 cases in the control) at the Second Xiangya Hospital, Central South University, China.

The intervention for the control group was group stretching and balance exercises over 12 weeks while the PMR group practiced relaxation of the muscles based on the Bernstein and Borkovec approach. The muscles were inclusive of "*muscles of the right hand and forearm, right biceps, left hand and forearm, left biceps, forearm, upper section of cheeks and nose, lower section of cheeks and nose, neck and throat, chest, shoulders and upper part of back, abdominal region and stomach, right thigh, right calf, right foot, left thigh, left calf, and left foot*".

The mental component summary score (QOL-MCS) of the quality of life (QOL) and depression scores, as evaluated by the Hospital Anxiety and Depression Scale (HADS) in PAH patients showed an improvement in the PMR group that was not seen in the control (Li et al, 2015).

The results of the promise in employing PMR on depression in hypertension patients are demonstrated below in table 14: 1, 2 and 3 (Copyright © 2015 Yunping Li et al. *This is an open access article distributed under the Creative Commons Attribution License, which permits unrestricted use, distribution, and reproduction in any medium, provided the original work is properly cited*):

4.9 BASELINE VS. POST-PMR SCORES FOR ANXIETY, DEPRESSION, AND QUALITY OF LIFE (QOL) WITH SIX-MINUTE WALK DISTANCE

T1: baseline; T2: 12 weeks after baseline

	Control (n=59)				PMR (n=55)			
	T1	T2			T1	T2		
Anxiety	7.4 ± 3.6	6.9 ± 3.2	−0.5 ± 3.4	.51	7.3 ± 3.5	4.9 ± 2.2	−2.4 ± 2.9[a]	<.01
Depression	7.2 ± 3.8	6.8 ± 3.4	−0.4 ± 3.5	.53	7.1 ± 3.5	4.8 ± 2.4	−2.3 ± 3.1[a]	<.01
QOL-PCS	46.2 ± 20.2	50.2 ± 24.0	4.0 ± 1.9	.38	46.3 ± 21.8	52.5 ± 23.2	6.2 ± 2.0	.17
QOL-MCS	54.8 ± 25.0	59.3 ± 25.2	4.5 ± 2.1	.38	54.9 ± 25.1	69.0 ± 26.3	14.1 ± 4.5[a]	<.01
Overall QOL	50.5 ± 22.62	54.7 ± 24.5	4.2 ± 2.0	.34	50.6 ± 23.6	60.8 ± 24.8	10.2 ± 3.7[a]	.03
6MWD	374.9 ± 136.2	405.6 ± 130.5	30.7 ± 9.5	.40	368.2 ± 143.5	402.5 ± 142.3	34.3 ± 10.6	.21

4.9.1. Depression and Anxiety

	Control				PMR		
		T1	T2		T1	T2	
HADS-anxiety	≤8	17 (29)	21 (36)	0.56	18 (33)	40 (73)	<0.01
	>8	42 (71)	38 (64)		37 (67)	15 (27)	

HADS-depression	≤8	22 (37)	25 (42)	0.71	24 (44)	38 (69)	0.01
	>8	37 (63)	34 (58)		31 (56)	17 (31).	

4.9.2 Quality of Life (QOL)

QOL domains	Control				PMR			
	T1	T2			T1	T2		
Physical function	38.3 ± 21.2	41.9 ± 23.2	3.6 ± 1.8	.38	37.9 ± 21.2	42.1 ± 23.0	4.2 ± 1.9	.35
Physical role	37.5 ± 20.7	42.7 ± 24.1	5.2 ± 2.6	.23	37.1 ± 20.5	42.9 ± 22.8	5.8 ± 2.7	.15
Body pain	69.1 ± 26.8	73.3 ± 29.1	4.2 ± 1.9	.42	69.4 ± 27.1	74.1 ± 29.3	4.7 ± 2.4	.63
General health	39.9 ± 18.4	43.0 ± 19.1	3.1 ± 1.5	.37	40.6 ± 19.3	42.9 ± 18.6	2.3 ± 1.1	.51
Vitality	43.3 ± 16.9	47.3 ± 17.3	4.0 ± 1.9	.22	42.8 ± 15.8	54.0 ± 22.5	11.2 ± 3.3[a]	<.01
Social function	59.8 ± 25.3	65.5 ± 25.6	5.7 ± 2.4	.26	60.5 ± 24.9	75.6 ± 27.9	15.1 ± 5.5[a]	<.01
Emotional role	60.3 ± 33.5	66.1 ± 35.2	5.8 ± 2.9	.38	59.1 ± 34.2	74.9 ± 28.0	15.8 ± 6.1[a]	<.01
Mental health	55.7 ± 24.3	58.2 ± 24.0	2.5 ± 1.1	.59	57.3 ± 25.3	71.7 ± 28.3	14.4 ± 4.6[a]	<.01

Given the extensive usage of mobile phones, Paulo Menezes and team (2019) reported the utility of *Emotional Control* (CONEMO) an app described as "a low-intensity psychoeducational 6-week intervention delivered via mobile phones assisted by a nurse" aimed at addressing depression in people with hypertension or diabetes. There were 3 pilot studies: one in São Paulo (the largest city of Brazil with 10% significant depression prevalence) and two in Lima, Peru (17.5% depression prevalence). Depression was assessed via the Patient Health Questionnaire-9 (PHQ-9) in 66 participants (41-60 years old). The app was inclusive of 3 sessions/week on motivation, self-care and boosting one to participate in healthy tasks. The nurses under supervision by research clinical psychologists were involved in initial contact, introducing the CONEMO app and monitoring.

The app screenshots are presented below in figure 6 (©Paulo Menezes, Julieta Quayle, Heloísa Garcia Claro, Simone da Silva, Lena R Brandt, Francisco Diez-Canseco, J Jaime Miranda, LeShawndra N Price, David C Mohr, Ricardo Araya. Originally published in JMIR Mental Health (http://mental.jmir.org), 26.04.2019. *This is an open-access article distributed under the terms of the Creative Commons Attribution License (https://creativecommons.org/licenses/by/4.0/), which permits unrestricted use, distribution, and reproduction in any medium, provided the original work, first published in JMIR Mental Health, is properly cited. The complete bibliographic information, a link to the original publication on http://mental.jmir.org/, as well as this copyright and license information must be included*):

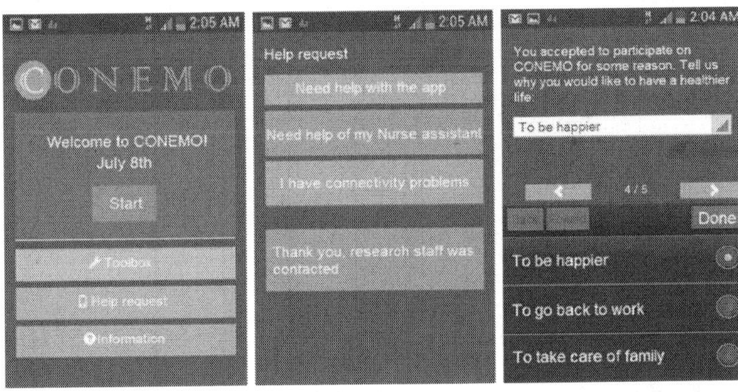

30% of São Paulo participants showed a mild risk to suicide at baseline that was not seen in Lima participants: this risk was not seen at the follow-up assessments. Further, the recovery from depression as assessed by PHQ-9 <10 was 65% in São Paulo, 87% in the first Lima pilot study and 75% in the second Lima pilot study. The levels of functional disability also showed a decline towards the end of the assessment. This trend is presented below in figure 7 (©Paulo Menezes, Julieta Quayle, Heloísa Garcia Claro, Simone da Silva, Lena R Brandt, Francisco Diez-Canseco, J Jaime Miranda, LeShawndra N Price, David C Mohr, Ricardo Araya. Originally published in JMIR Mental Health (http://mental.jmir.org), 26.04.2019. *This is an open-access article distributed under the terms of the Creative Commons Attribution License (https://creativecommons.org/licenses/by/4.0/), which permits unrestricted use, distribution, and reproduction in any medium, provided the original work, first published in JMIR Mental Health, is*

properly cited. The complete bibliographic information, a link to the original publication on http://mental.jmir.org/, as well as this copyright and license information must be included):

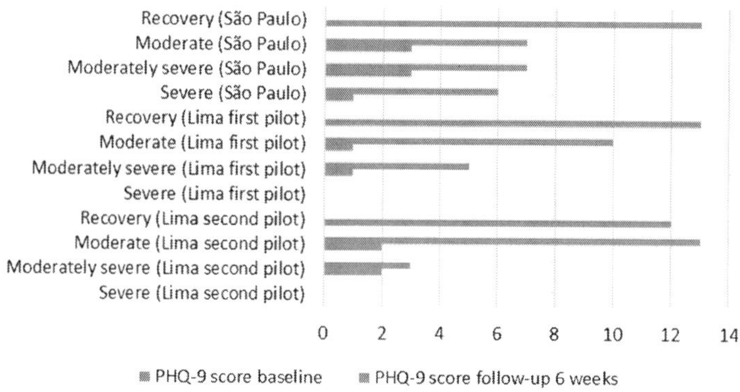

Table 15 below highlights the changes in quality of life and disability following assessment (©Paulo Menezes, Julieta Quayle, Heloísa Garcia Claro, Simone da Silva, Lena R Brandt, Francisco Diez-Canseco, J Jaime Miranda, LeShawndra N Price, David C Mohr, Ricardo Araya. Originally published in JMIR Mental Health (http://mental.jmir.org), 26.04.2019. *This is an open-access article distributed under the terms of the Creative Commons Attribution License (https://creativecommons.org/licenses/by/4.0/), which permits unrestricted use, distribution, and reproduction in any medium, provided the original work, first published in JMIR Mental Health, is properly cited. The complete bibliographic information, a link to the original publication on http://mental.jmir.org/, as well as this copyright and license information must be included*):

Scales and categories Baseline	São Paulo (N=20), n (%)		Lima—pilot 1 (N=15), n (%)		Lima—pilot 2 (N=16), n (%)	
	6-week follow-up	Baseline	6-week follow-up	Baseline	6-week follow-up	
World Health Organization Disability Assessment Schedule						
No disability	0 (0)	2 (10)	0 (0)	1 (7)	0 (0)	0 (0)
Mild disability	9 (45)	10 (50)	5 (33)	9 (60)	7 (44)	9 (56)
Moderate disability	8 (40)	5 (25)	7 (47)	5 (33)	7 (44)	7 (44)
Severe disability	3 (15)	3 (15)	2 (13)	0 (0)	2 (13)	0 (0)

EuroQoL (Quality of Life)-5 dimensions						
Problems to walk	10 (50)	11 (55)	10 (67)	7 (47)	7 (44)	5 (31)
Problems washing or dressing	9 (45)	5 (25)	3 (20)	2 (13)	1 (6)	0 (0)
Problems with performing usual activities	9 (45)	9 (45)	8 (53)	4 (27)	5 (31)	5 (31)
Moderate or extreme pain/ discomfort	14 (70)	16 (80)	14 (93)	11 (73)	13 (81)	11 (69)
Moderate or extreme anxiety/ depression	17 (85)	15 (75)	7 (47)	7 (47)	12 (75)	8 (50)

The feedback of using an "app" to control depression associated with hypertension or diabetes is presented below in table 16 (©Paulo Menezes, Julieta Quayle, Heloísa Garcia Claro, Simone da Silva, Lena R Brandt, Francisco Diez-Canseco, J Jaime Miranda, LeShawndra N Price, David C Mohr, Ricardo Araya. Originally published in JMIR Mental Health (http:// mental.jmir.org), 26.04.2019. *This is an open-access article distributed under the terms of the Creative Commons Attribution License (https:// creativecommons.org/licenses/by/4.0/), which permits unrestricted use, distribution, and reproduction in any medium, provided the original work, first published in JMIR Mental Health, is properly cited. The complete bibliographic information, a link to the original publication on http:// mental.jmir.org/, as well as this copyright and license information must be included*):

TOP: Participant evaluation

Intervention aspects evaluated	São Paulo (average grade)	Lima pilot study 1 (average grade)	Lima pilot study 2 (average grade)
Helped with physical health	4.1	4.0	3.8
Helped with mental health	4.4	4.4	4.3
Helped having the will to do things	3.9	4.3	4.3

Helped to get organized	3.6	4.3	4.3
Count on nurse assistant or nurse	4.5	4.1	4.1
Nurse assistant or nurse helped	4.4	4.3	3.8
Number of contacts	4.1	3.6	3.2
Training quality	4.1	3.7	3.1
Intervention duration	4.1	2.8	2.9
Achievement of objectives	4.1	4.2	4.2
Satisfaction with results	4.3	4.4	4.3
Would indicate to a friend	3.6	4.6	4.5

BOTTOM: Technological feedback:

Technological aspect	**São Paulo (average grade)**	**Lima pilot study 1 (average grade)**	**Lima pilot study 2 (average grade)**
Easy to use	4.4	4.3	3.8
Usefulness	4.6	4.7	4.4
Adequacy of frequency of sessions	4.3	3.3	3.9
Content interesting	4.4	4.6	4.5
Video quality	4.4	4.6	4.4
Sound quality	3.9	4.6	4.3
Choices presented	4.1	4.6	4.5
Training session	4.3	4.0	4.3
Reminders	3.6	3.0	4.3

Overall the researchers concluded that the lowering of severity of depression and the positive feedback from participants open up the avenue of using such an approach to help quell depressive symptoms in hypertensive people.

A randomized control trial (RCT) by Peterson and team (2018) has been put forth in *Trials* described "*as a pragmatic, two-arm, parallel-cluster RCT, the unit of randomisation being the clinics, with outcomes being measured for individual participants*" is aimed at targeting depression in hypertensive patients at primary health care (PHC) facilities in South Africa (SA).

The proposed approach looks at identifying depression in hypertensive patients on the basis of the Patient Health Questionnaire (PHQ-9) from 20 largest eligible clinics from one district in the North West Province. Training on Primary Care 101 (PC101) that a clinical decision support tool is planned for the nurse clinicians. Appropriate counseling by a behavioral health counselor allows for bolstered referral systems. The definition of the primary outcome is at least 50% lowering in the PHQ-9 score at 6 months: South African National Clinical Trial Register: SANCTR (http://www.sanctr.gov.za/SAClinicalTrials) (DOH-27-0916-5051) as presented in the scheme in figure 8 below (Petersen et al, 2018: *Open Access: This article is distributed under the terms of the Creative Commons Attribution 4.0 International License (http://creativecommons.org/licenses/by/4.0/), which permits unrestricted use, distribution, and reproduction in any medium, provided you give appropriate credit to the original author(s) and the source, provide a link to the Creative Commons license, and indicate if changes were made. The Creative Commons Public Domain Dedication waiver (http://creativecommons.org/publicdomain/zero/1.0/) applies to the data made available in this article, unless otherwise stated*):

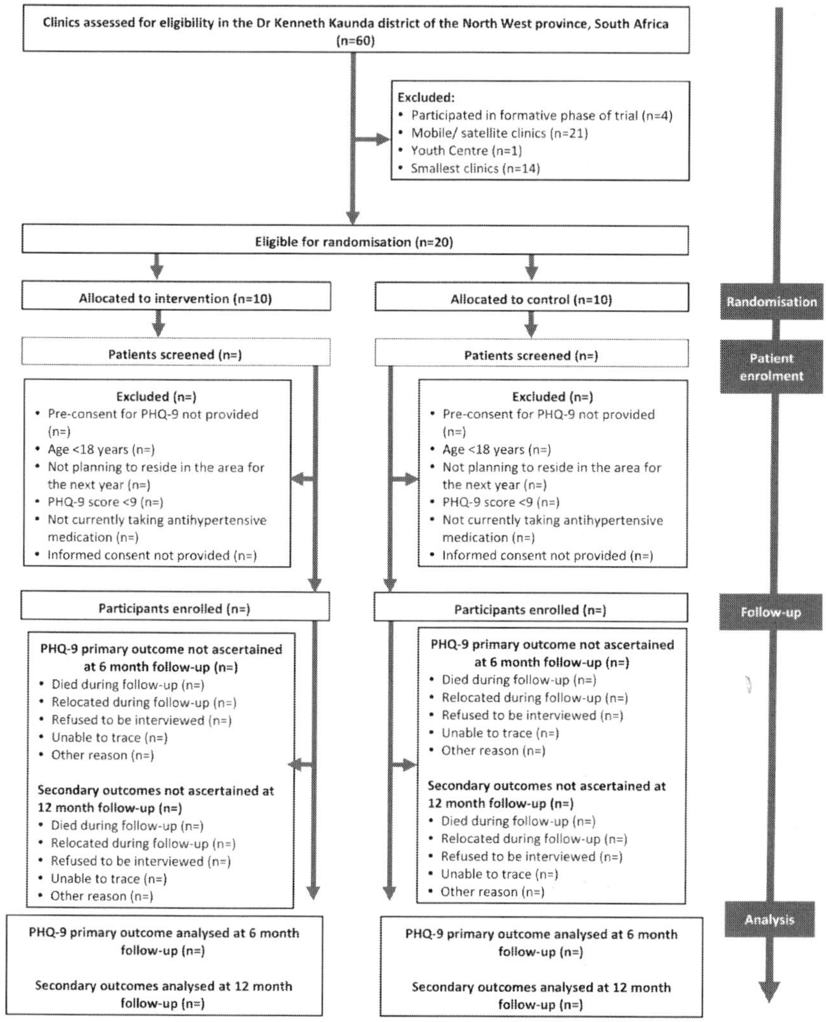

The scheme of managing depression in hypertensive patients has been presented below in figure 9 (Petersen et al, 2018: *Open Access: This article is distributed under the terms of the Creative Commons Attribution 4.0 International License (http://creativecommons.org/licenses/by/4.0/), which permits unrestricted use, distribution, and reproduction in any medium, provided you give appropriate credit to the original author(s) and the source, provide a link to the Creative Commons license, and indicate if changes were made. The Creative Commons Public Domain Dedication waiver (http://creativecommons.org/publicdomain/zero/1.0/) applies to the data made available in this article, unless otherwise stated*):

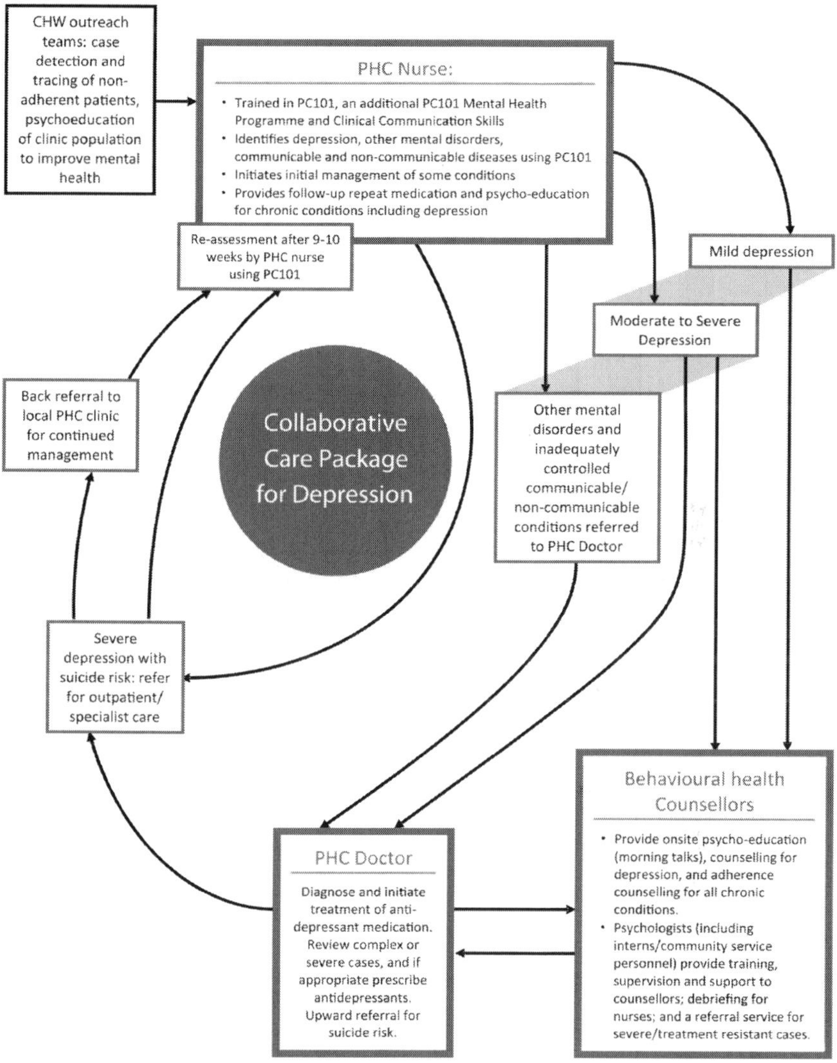

As part of the intervention, the PC101 guidelines by the researchers have been presented below (Petersen et al, 2018: *Open Access: This article is distributed under the terms of the Creative Commons Attribution 4.0 International License (http://creativecommons.org/licenses/by/4.0/), which permits unrestricted use, distribution, and reproduction in any medium, provided you give appropriate credit to the original author(s) and the source, provide a link to the Creative Commons license, and indicate if changes were made. The Creative Commons Public Domain Dedication waiver (http://creativecommons.org/publicdomain/zero/1.0/) applies to the data made available in this article, unless otherwise stated*):

" 1. *In the waiting room, the Recruiter gives the following information to the group of patients assembled there. A suggested script is provided*:

"My name is xx and I am employed by the University of Cape Town to work on two studies investigating whether counselling helps improve health outcomes for people with chronic conditions. We are looking for clinic patients over the age of 18 years with certain chronic diseases to take part. These studies will help us to improve the treatment chronic patients receive at clinics in South Africa.The studies will involve being interviewed today and twice more, in 6 months' time and in 12 months' time.For now, we are asking you to agree to being asked three short questions to see if you might be suitable for the studies. We will ask you some more detailed questions to determine whether you qualify to take part in either or both of the studies. Should you qualify we will also give you more detailed information about the full studies so that you can decide whether or not to take part.Your care at the clinic will not be affected in any way if you choose not to take part. But we hope you will take part. I will be based [state where the recruiter will be based e.g. in the waiting room or interview room] so please come and tell me if you are interested in taking part in these important studies"

1. *The Recruiter engages with interested patients, one on one, either in a quiet corner of the waiting room or preferably in a private space close to where the interviews are being conducted. He/ she asks the interested patient the following 3 questions:*

   "Thank you for your interest in the study. We are looking for patients on high blood pressure treatment and/or antiretrovirals who plan to stay in the local area for the duration of the study. Can I therefore ask:

   i. *Are you currently taking medicine for high blood pressure (hypertension or high blood)?*
   ii. *Are you currently taking ARVs (antiretrovirals)?*
   iii. *If yes to [i] and/or [ii] ask:*
   iv. *Are you planning to stay in the area for the next year?*

   If yes to [i] and/or [ii] AND yes to [iii]:

   "It looks like you are suitable to be screened for participation in the study. My colleague who is completing the full interviews will screen you using some questions. Are you still interested in taking part?"

   If no to these questions, thank the patient for their time, remind them their care at the clinic will not be affected because they didn't qualify for the study and wish them well.

2. *Recruiter arranges for Interviewer to screen potential participant.*

3. *The Interviewer now engages with the potential participant in a private space. He/ she starts by asking in what language the potential participant would like to be interviewed, selecting the appropriate interview option on the handheld device, and capturing the date and clinic in which the screen is being done. No patient identifiers are captured.*

4. *(Built into electronic questionnaire.) The interviewer then proceed with the electronic questionnaire which includes the following opt-out consent to be screened using PHQ-9:*

   "As you have heard from my assistant we are conducting two studies to evaluate a programme and we are looking for people with certain criteria to take part. I would like to start by asking you a few questions to see whether you qualify to take part in either or both of the studies. Some of these questions are about your psychological well-being and may be sensitive. You are free to stop the interview at any time and this will not affect your usual care at the clinic. Are you willing to continue with the questionnaire?"

   If no, Interviewer to read: 'We would like to thank you for your time today and we wish you well.'

   If yes, continue.

5. *(Built into electronic questionnaire.) Continue electronic questionnaire to PHQ-9 and three questions on whether or not the patient is on antihypertensives or antidepressants and planning to stay in the vicinity of the clinic for the study period.*

   If patient answers that they have had end-of-life thoughts in response to PHQ-9 Question 9 (i.e. 1-7 days, 8-11 days, 12-14 days), repeat that question (built into questionnaire programme).

If 1-7 days, Interviewer to read: 'At the end of the interview I will give you a leaflet with some contact details for people who will be able to help you with how you are feeling.'

If 7 days or more, Interviewer to read: 'I am concerned about how you are feeling and would like to ask a nurse to see you at the end of this interview.'

If not eligible: Skip to end of questionnaire.
If eligible, continue.
Note: No confidential information (full names, date of birth, contact details) may be recorded for patients who have not yet provided written informed consent.

6. *Interviewer to go through written consent process for eligible patients. Interviewer goes through patient information sheets for one or both studies, explaining study procedures, potential risks and benefits and asks patient whether or not they are still willing to participate.*

   If yes, the Interviewer continues with the interview, now capturing patients' confidential information.

   If no, continue to 8.

7. *(Built into electronic questionnaire.) Interviewer to review patient/participant's status in terms of end-of-life thoughts, as prompted by computer.*

   If the patient confirmed end-of-life thoughts further actions to be guided by frequency of those thoughts.

   Less than 7 days in the last 2 weeks: Provide patient with written material on depression and suicide prevention.

   7 days or more in the last 2 weeks: Complete urgent referral form, and accompany patient to clinic staff. Do not leave patient unattended. Wait until patient has been handed over to clinic staff" (Peterson et al, 2018).

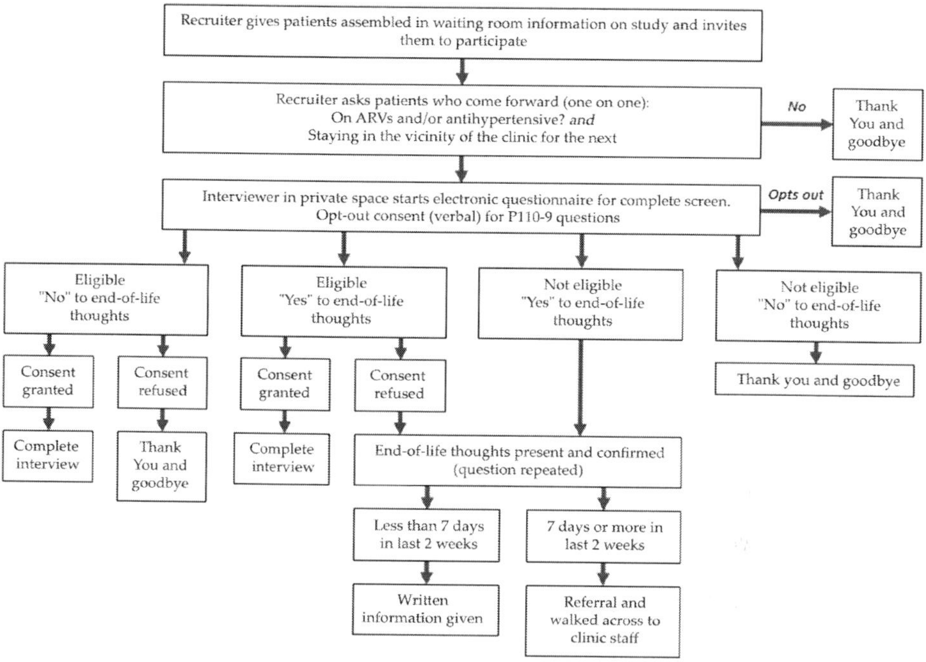

Further, the scheme of interventions and their plans to address depression in hypertension are depicted below in table 17 (Petersen et al, 2018: *Open Access: This article is distributed under the terms of the Creative Commons Attribution 4.0 International License (http://creativecommons.org/licenses/by/4.0/), which permits unrestricted use, distribution, and reproduction in any medium, provided you give appropriate credit to the original author(s) and the source, provide a link to the Creative Commons license, and indicate if changes were made. The Creative Commons Public Domain Dedication waiver (http://creativecommons.org/publicdomain/zero/1.0/) applies to the data made available in this article, unless otherwise stated*):

Key: NCD noncommunicable diseases, *PHC* primary health care, *WHO* World Health Organisation

Provider	**Role**	**Training**	**Content of training**	**Method and time-frame**
Control and intervention facilities				
PHC nurses	Identifies, provides brief interventions and refers	Basic on-site PC101 training	Case scenarios used for training in the identification and management of common chronic diseases, including communicable diseases, NCDs (including hypertension), women's health and mental health. Mental health components draw on the WHO's mhGAP guidelines [67] and adopt a syndromic approach to mental health symptoms (such as stress, insomnia, suicidal thinking) with diagnostic algorithms and treatment checklists for depression	(1) PC101 master trainers train facility trainers who train PHC nurses at the facilities
				(2) 12 weekly sessions over 12 weeks at facilities (2 of which are on mental disorders)
				(3) Training uses case-scenario material of patients with chronic conditions, including co-morbid conditions
Intervention facilities				

PHC nurses	Identifies, provides brief interventions and refers	Orientation and clinical communication skills training	(1) Overview of the system changes being made by the DoH in South Africa to accommodate the demands of integrated chronic care; their role as case managers within the collaborative-care model for depression	4 2-h interactive workshops at PHC facilities/regional training centre
			(2) Orientation to patient-centred care and clinical communication skills necessary to implement patient-centred care	
			(3) Skills to manage patient emotions within the consultation; self-care including how to cope with their own emotions and burn-out	
			(4) Motivational interviewing skills to promote patient self-management	
		PC101 supplementary training in mental health	(1) Detection of depression and anxiety, psychoeducation and referral to counsellors and/or physician for consideration of psychotropic medication in the case of moderate to severe depression	(1) PC101 master trainers train facility trainers (2-day workshop) who train PHC nurses at the facilities

			(2) Detection of risky alcohol use and brief intervention for harmful/ hazardous drinking and for detoxification and referral to specialist rehabilitation programmes for dependency as per the mhGAP guidelines [68]	(2) 3 weekly sessions over 3 weeks at facilities, with an additional follow-up session 1 month later
			(3) Assessment of suicide intent	
			(4) Patient review after 8 weeks to assess response to treatment and onward referral for specialist care as indicated by the mhGAP evidence-based guidelines for LMICs [68] if necessary following a treatment-to-target approach as contained in the collaborative-care model (see Fig. 3). Treatment to target involves tracking a patient's symptom severity and adjusting or intensifying treatment should patients not show an improvement in symptoms following initial treatment [69]	(3) Training uses case scenarios case scenario material of chronic patients with co-morbid mental disorders

PHC physicians	Diagnoses, initiates and monitors response to psychotropic medication	Orientation and training in mhGAP/ PC101	(1) Orientation to the importance of treating co-morbid depression	3 1-day workshops spread over 6 months
			(2) Training in mhGAP guidelines	
			(3) Follow-up using case studies of patients	
Behavioural health counselors	Provides evidence-based counselling	Counselling training	(1) Manualised counselling package comprising 8 sessions (delivered individually or in groups)	1 week of off-site training; 1 week of peer-to-peer mentoring; in-vivo supervision by a psychologist of each session; weekly follow-up group supervisory sessions, augmented where possible by weekly individual supervision sessions

			(2) Session 1: psychoeducation session on depression; the last session is a closure session; sessions 2-7 draw on problem solving and cognitive behavioural techniques, including behavioural activation to address the common triggers of depression and anxiety which, in this population, include poverty, interpersonal conflict, social isolation and avoidance, grief and loss, and stigma that emerged from qualitative interviews held with service users with depression during the formative phase of the PRIME project in South Africa in 2 provinces [70]. A prototype had been field tested in KwaZulu-Natal and positive results demonstrated in an individually randomised pilot trial [42, 71]; adherence session provides information on the chronic condition/s and chronic medication/s the patients may have as well as helping patients with adherence difficulties	

			(3) While developed to treat depression, the intervention has been found to promote improvements in global psychological functioning as well [42, 71], thus having the potential for trans-diagnostic effects, in line with evidence that diagnosis-specific cognitive-behavioural therapy has beneficial effects on untargeted comorbid emotional disorders [72]	
Specialists (psychologist/psychiatrist)	Training, supervision of counsellors	Orientation to task sharing	Psychologists (including interns and community service psychologists) orientated to their roles	One-off workshops

4.10 DIABETES AND HYPERTENSION

The coexistence of hypertension and type 2 diabetes (T2D) has been highlighted by research with 70% of individuals with T2D getting affected by hypertension. Further, in those with preexisting hypertension, the development of T2D is higher by a factor of 2.5 (Gress et al, 2000). The risk of all-cause mortality is also increased when both these conditions are resent as opposed to either condition alone (Chen et al, 2011).

To show an example of how this combination affects the body, Bassi and team (2018) scrutinized the modulation of the cardiac autonomic nervous system in patients with hypertension (HTN) and T2DM. The RR (the time between two consecutive R signals in an ECG) intervals of 60 T2DM patients (28 also showing hypertension) from the cardiovascular outpatient clinic of the Federal University of Sao Carlos was done. Further quantification of the heart rate variability (HRV) was done in these patients. The tests revealed that the mean RR interval in T2DM +HTN patients was 801.1 ms compared to 871.5 ms in only T2DM patients. The linear and nonlinear HR dynamics were found to be targeted in the co-morbid group vs. that of only the diabetic group as seen with lowered non-linear indices SD1,

Shannon entropy and ApEn values. Interestingly, higher insulin resistance was detected in the HTN+ T2DM group vs. T2DM alone as seen by a higher profile of HOMA-IR: homeostatic model assessment insulin resistance and insulin in the former.

Table 18 below depicts the changes in the autonomic cardiac system in T2DM plus HTN vs. T2DM alone (Bassi et al, 2018:):

Variables	**DMT2 (n = 32)**	**DMT2+HTN (n = 28)**	**p value**
Linear			
Mean RR intervals (ms)	871.5 ± 105.8	801.1 ± 89.0*	0.010
RMSSD (ms)	17.9 ± 11.1	21.2 ± 15.2	0.358
STD RR	29.3 ± 21.5	31.5 ± 23.2	0.718
LF (nu)	66.3 ± 19.8	59.7 ± 22.9	0.247
HF (nu)	33.7 ± 19.8	40.3 ± 22.9	0.241
TINN	110.5 ± 59.8	121.3 ± 67.5	0.523
RR Tri	5.5 ± 2.6	7.1 ± 4.5	0.082
Nonlinear			
SD1	14.5 ± 8.2	9.5 ± 4.4*	0.021
SD2	40.4 ± 20.0	43.0 ± 23.1	0.662
SE	3.2 ± 0.3	3.0 ± 0.3*	0.012
ApEn	14.5 ± 8.2	9.5 ± 4.4*	0.021
SampEn	1.4 ± 0.3	1.5 ± 0.3	0.601
CD	1.2 ± 1.3	1.6 ± 1.6	0.271

A cross-sectional study of 109 diabetic (both type 1 and type 2) for risk factors of hypertension was done at the Yaoundé Central Hospital and the Etoug-Ebe Baptist Health Center of Cameroon was reported by Kemche and team (2020) in the Journal of Diabetes Research. Analyses revealed that the prevalence of hypertension in these diabetic patients was 86.2%. Further classification on the basis of 2017 ACC/AHA/AAPA/ABC/ACPM/AGS/APhA/ASH/ASPC/NMA/PCNA Guideline for the Prevention, Detection, Evaluation, and Management of High Blood Pressure in Adults: A Report of the American College of Cardiology/American Heart Association Task Force on Clinical Practice Guidelines revealed that 32.1% were stage 1 hypertensive and 54.1% were stage 2 hypertensive.

Analyses of the risk factors showed that gender, feeding frequency, duration of diabetes, and smoking were the risks for stage 1 hypertension. In the case of stage 2 hypertension, the risk factors were age, duration of diabetes, and waist circumference. The results are presented below in table 19 (Copyright © 2020 Beryl Kemche et al. *This is an open access article distributed under the Creative Commons Attribution License, which permits unrestricted use, distribution, and reproduction in any medium, provided the original work is properly cited*):

TOP: *Diabetic patients and hypertension: the same superscript is indicative of statistical significance*

Factors	Normotensive	Stage 1 hypertensive	Stage 2 hypertensive
Age (years)			
>40 years	11[a] (73.3%)	29[ab] (82.9%)	56[b] (94.9%)
≤40 years	4 (26.7%)	6 (17.1%)	3 (5.1%)
Statistics	;		
Gender			
Female	35 (36.4%)	19 (54.3%)	12 (80.0%)
Male	24 (63.6%)	16 (45.7%)	3 (20.0%)
Statistics	;		

BOTTOM: The risk factors for hypertension stages in diabetics in this study: *Key: OR: odds ratio; 95% CI: 95% confidence interval; n:frequency*

Predictor variables	Hypertensive stage 1 (35)	Normotensive (15)	Univariate analysis		Multivariate analysis	
	(%)	(%)	OR (95% CI)	value	OR (95% CI)	value

Gender						
Male	16 (45.7%)	3 (20.0%)	1 (reference)		0.058 (0.004-0.956)	0.046
Female	19 (54.3%)	12 (80.0%)	0.297 (0.071-1.240)	0.086	1 (reference)	
Feeding rate						
≥3 times a day	23 (65.7%)	7 (46.7%)	2.190 (0.639-7.504)	0.208	6.478 (1.035-40.533)	0.046
<3 times a day	12 (34.3%)	8 (53.3%)	1 (reference)		1 (reference)	
Duration of diabetes						
≥10 years	17 (48.6%)	4 (26.7%)	2.597 (0.692-9.747)	0.150	0.856 (0.739-0.992)	0.039
<10 years	18 (51.4%)	11 (73.3%)	1 (reference)		1 (reference)	
Smoking status						
Yes	2 (5.7%)	3 (20.0%)	0.242 (0.036-1.633)	0.123	0.017 (0.000-0.769)	0.036
No	33 (94.3%)	12 (80.0%)	1 (reference)		1 (reference)	
Occupation						
Not retired or not a housewife	21 (60.0%)	5 (33.3%)	1 (reference)		0.435 (0.080-2.361)	0.335
Retired or housewife	14 (40.0%)	10 (66.7%)	0.333 (0.094-1.185)	0.084	1 (reference)	
Tribe						
Not from the west region	18 (51.4%)	8 (53.3%)	1.079 (0.321-3.626)	0.902	0.257 (0.039-1.699)	0.159
West region	17 (48.6%)	7 (46.7%)	1 (reference)		1 (reference)	

From stage 1 to stage 2 in hypertension	Hypertensive stage 2 (59)	Hypertensive stage 1 (35)				
Age						
>40 years	56 (94.9%)	29 (82.9%)	3.862 (0.900-16.574)	0.055	0.916 (0.866-0.969)	0.002
≤40 years	3 (5.1%)	6 (17.1%)	1 (reference)		1 (reference)	
Duration of diabetes						
≥10 years	13 (22.0%)	17 (48.6%)	0.299 (0.121-0.739)	0.008	1.155 (1.051-1.270)	0.003
<10 years	46 (78.0%)	18 (52.4%)	1 (reference)		1 (reference)	
Waist circumference						
Abnormal	55 (93.2%)	27 (77.1%)	4.074 (1.126-14.735)	0.024	1.005 (0.965-1.047)	0.802
Normal	4 (6.8%)	8 (22.9%)	1 (reference)		1 (reference)	
Normotensives to hypertensives	Hypertensive (94)	Normotensive (15)				
Waist-to-hip ratio						
Abnormal	83 (88.3%)	10 (66.7%)	3.773 (1.087-13.091)	0.028	0.000 (0.000-0.007)	0.003
Normal	11 (11.7%)	5 (33.3%)	1 (reference)		1 (reference)	

Smoking status						
Yes	5 (5.3%)	3 (20.0%)	0.225 (0.048-1.062)	0.043	0.175 (0.031-1.008)	0.051
No	89 (94.7%)	12 (80.0%)	1 (reference)		1 (reference)	
Feeding rate						
≥3 times a day	66 (70.2%)	7 (46.7%)	2.694 (0.891-8.146)	0.072	3.417 (1.019-11.455)	0.046
<3 times a day	28 (29.8%)	8 (53.3%)	1 (reference)		1 (reference)	

Overall, lack of controlled BP could be predicted by smoking, higher age, long time span of hypertension, lack of adherence to medication and uncontrolled glycaemia (Kemche et al, 2020).

2019-published research in *Diabetology & Metabolic Syndrome* by Chang-Yuan Liu and team looked at the clinical aspects of newly diagnosed hypertension in diabetes mellitus and newly diagnosed diabetes mellitus in hypertension. The patient set was from a cross-sectional study at cardiovascular and endocrine departments of hospitals between June 2011 and March 2012 called the China multicentre registry. Measurement of blood pressure, electrocardiogram (ECG), urine routine and other aspects were done in 137 hypertensive patients with newly diagnosed diabetes mellitus (measured by an oral glucose tolerance test: OGTT) and 155 diabetic patients with newly diagnosed hypertension.

A lower profile of lower plasma fasting glucose and HbA1c was observed for hypertensive patients with newly diagnosed diabetes mellitus, compared to diabetic patients with newly diagnosed hypertension. An increase of body mass index and heart rate was reported from 40% to 45.8% for regular weight and 40% to 58.1% in obese hypertensive patients with newly diagnosed diabetes mellitus. The extent of ischaemic heart disease was more while that of albuminuria was lower evidently in hypertensive patients with newly diagnosed diabetes mellitus vs. diabetic patients with newly diagnosed hypertension.

Overall, the increased BMI and left ventricular hypertrophy and ischaemic heart disease in hypertensive patients with newly diagnosed diabetes mellitus with a lower extent of albuminuria compared to diabetic patients with newly diagnosed hypertension was reported to be suggestive of the different patterns of organ damage depending on which condition strikes first.

Figure 10 below is illustrative of the changes in hypertensive patients with newly diagnosed diabetes mellitus as opposed to diabetic patients with newly diagnosed hypertension (Liu et al, 2019: *Open Access: This article is distributed under the terms of the Creative Commons Attribution 4.0 International License (http://creativecommons.org/licenses/by/4.0/), which permits unrestricted use, distribution, and reproduction in any medium, provided you give appropriate credit to the original author(s) and the source, provide a link to the Creative Commons license, and indicate if changes were made. The Creative Commons Public Domain Dedication waiver (http://creativecommons.org/publicdomain/zero/1.0/) applies to the data made available in this article, unless otherwise stated*):

TOP: body mass index (**a**) and heart rate (**b**)

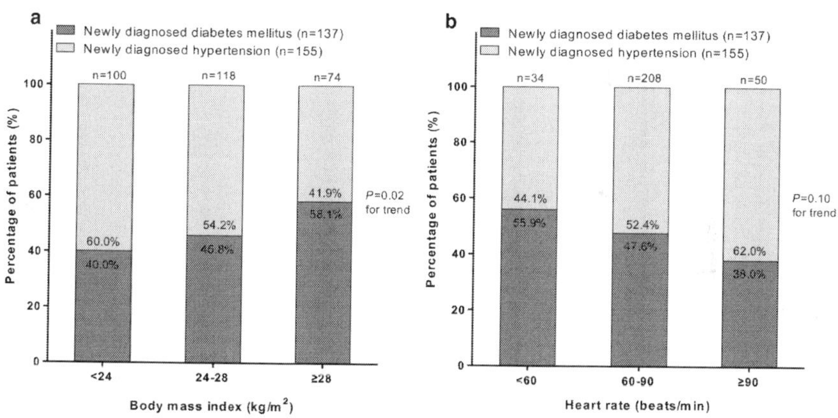

BOTTOM: albuminuria and electrocardiographic (ECG)-left ventricular hypertrophy

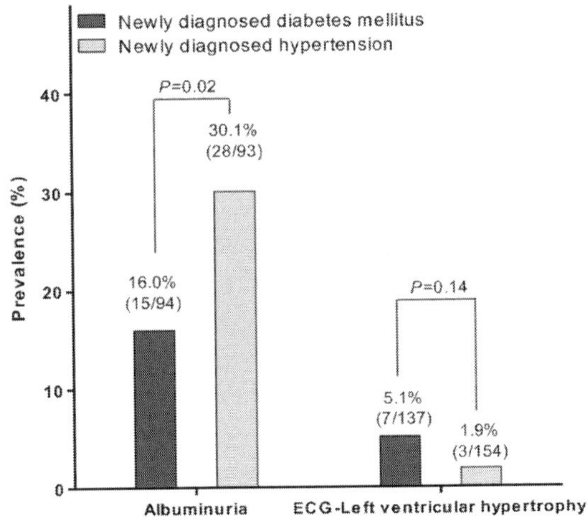

In order to **address these two conditions**, a combination of lifestyle modifications, awareness, telemedicine, boosted infrastructure at healthcare facilities along optimal public-private mergers to provide relevant healthcare facilities. Table 20 below outlines a **potential system** as proposed in a review in the International Journal of Hypertension (Copyright © 2013 Viswanathan Mohan et al. *This is an open access article distributed under the Creative Commons Attribution License, which permits unrestricted use, distribution, and reproduction in any medium, provided the original work is properly cited*):

	Detection	Diabetes education	Equipment	Medicines
Primary care	Large scale screening using high risk category approach	Through community workers—training of community workers in screening activities	Basic equipment like Glucometer, BP apparatus	Essential low cost generic drugs to be made available at all PHCs

Secondary care	Confirmation of diagnosis Screening for complications of diabetes and co-morbid conditions	(i) Physicians (ii) Paramedicals	Ophthalmology Basic laboratory support	Insulin, oral hypoglycemic drugs, ACE inhibitors, calcium antagonists, ARBs, statins, aspirin, and other antihypertensive drugs
Tertiary care	Treatment of complications	(i) Diabetologists (ii) Ophthalmologists (iii) Nephrologists (iv) Cardiologists	Advanced equipment (i) Laser photocoagulation (ii) Fundus camera (iii) Cardiac care (iv) Dialysis unit (v) Transplantation	All of above and costlier drugs or treatments

In a descriptive qualitative study of focus group discussions by Ndejjo and team (2021), challenges to addressing the chronic conditions and their suggested solutions to overcome the limitations are presented. They are presented by the researchers below in table 21 (© 2021 Ndejjo R *et al. This is an open access article distributed under the terms of the Creative Commons Attribution License, which permits unrestricted use, distribution, and reproduction in any medium, provided the original work is properly cited*):

Themes	Sub theme – challenges	Sub theme - coping
Getting diagnosed with hypertension and diabetes	Inadequate opportunities for diagnosis	Taking advantage of any opportunities to screen
Accessing care for hypertension and diabetes	High transportation costs to reach health facilities	Borrowing transport funds from colleagues and sometimes trekking to facilities
	Absence of health workers from stations resulting in long waiting time	Keeping scheduled appointments and early morning queuing in order not to miss health workers
	Inadequate information provided by health workers and their lack of follow-up	Obtain information from friends, traditional practitioners and media, and sharing experiences with peers
	Inadequate screening facilities and supplies	Forming support group to mobilise resources to purchase basic equipment and supplies
Accessing medications	Frequent drug stock-outs at public health facilities	Purchase of own drugs or access drugs through friends and social networks or skipping medications
	Expensive drugs at private-for-profit facilities	Obtain subsidized drugs from private-not-for-profit facilities or resorting to cheaper herbal remedies used alone or concurrently with conventional medicine
Monitoring and managing conditions	Inadequate knowledge about their conditions	Keenly obtain any information from health workers and sometimes peers to support self-management
	Few opportunities for self-monitoring conditions	Utilizing private facilities or community health workers for monitoring or purchasing required supplies through contributions by their patient's groups
	Lack of reminders for taking medication affecting adherence	Peers and family members acting as reminders to take medication.
Adopting lifestyle recommendations	Changing behaviors and adopting lifestyle recommendations was challenging	Fitting lifestyle changes within usual routine and mobilizing family members to also adopt changes while ignoring 'unrealistic' recommendations

4.11 FINAL POINTS

Given the challenge of hypertension and its associated impacts that is being uncovered, the screening and management of risk factors becomes vital. A combination of individual changes with community support and political impact can help timely healthcare and nip the trouble in the bud. Figure 11 below is illustrative of targeting hypertension and associated chronic ailments (Copyright © 2013 Viswanathan Mohan et al. *This is an open access article distributed under the Creative Commons Attribution License, which permits unrestricted use, distribution, and reproduction in any medium, provided the original work is properly cited*):

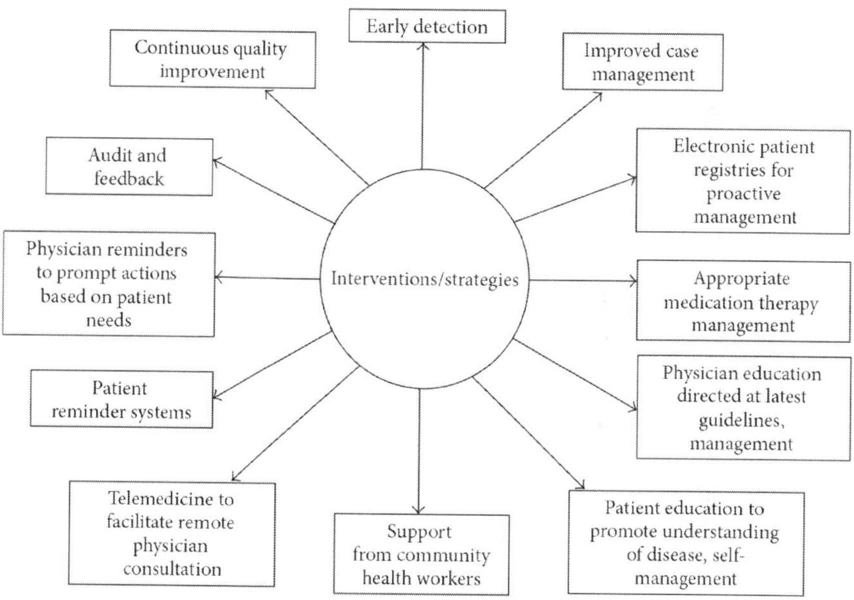

4.12 REFERENCES

1. Yasar, S., Ko, J. Y., Nothelle, S., Mielke, M. M., & Carlson, M. C. (2011). Evaluation of the effect of systolic blood pressure and pulse pressure on cognitive function: the Women's Health and Aging Study II. *PloS one*, *6*(12), e27976. https://doi.org/10.1371/journal.pone.0027976
2. Suzuki H, Gao H, Bai W, Evangelou E, Glocker B, O'Regan DP, et al. (2017) Abnormal brain white matter microstructure is associated with both pre-hypertension and hypertension. PLoS ONE 12(11): e0187600. https://doi.org/10.1371/journal.pone.0187600
3. Kurosaki H, Nakahata K, Donishi T, Shiro M, Ino K, Terada M, et al. (2018) Effects of perinatal blood pressure on maternal brain functional connectivity. PLoS ONE 13(8): e0203067. https://doi.org/10.1371/journal.pone.0203067
4. Bu L, Huo C, Xu G, Liu Y, Li Z, Fan Y and Li J (2018) Alteration in Brain Functional and Effective Connectivity in Subjects With Hypertension. *Front. Physiol.* 9:669. doi: 10.3389/fphys.2018.00669
5. Hestad K, Engedal K, Schirmer H and Strand BH (2020) The Effect of Blood Pressure on Cognitive Performance. An 8-Year Follow-Up of the Tromsø Study, Comprising People Aged 45–74 Years. *Front. Psychol.* 11:607. doi: 10.3389/fpsyg.2020.00607
6. Sun, D., Thomas, E.A., Launer, L.J. *et al.* Association of blood pressure with cognitive function at midlife: a Mendelian randomization study. *BMC Med Genomics* 13, 121 (2020). https://doi.org/10.1186/s12920-020-00769-y
7. Mansukhani MP, Kolla BP and Somers VK (2019) Hypertension and Cognitive Decline: Implications of Obstructive Sleep Apnea. *Front. Cardiovasc. Med.* 6:96. doi: 10.3389/fcvm.2019.00096
8. Roberts L, Davis GK and Homer CSE (2019) Depression, Anxiety, and Post-traumatic Stress Disorder Following a Hypertensive Disorder of Pregnancy: A Narrative Literature Review. *Front. Cardiovasc. Med.* 6:147. doi: 10.3389/fcvm.2019.00147
9. White, J., Hopkins, R.O., Glissmeyer, E.W. *et al.* Cognitive, emotional, and quality of life outcomes in patients with pulmonary arterial hypertension. *Respir Res* **7**, 55 (2006). https://doi.org/10.1186/1465-9921-7-55
10. Grimsrud A, Stein DJ, Seedat S, Williams D, Myer L (2009) The Association between Hypertension and Depression and Anxiety

Disorders: Results from a Nationally-Representative Sample of South African Adults. PLoS ONE 4(5): e5552. https://doi.org/10.1371/journal.pone.0005552

11. Hammami, S., Mehri, S., Hajem, S. *et al.* Awareness, treatment and control of hypertension among the elderly living in their home in Tunisia. *BMC Cardiovasc Disord* **11,** 65 (2011). https://doi.org/10.1186/1471-2261-11-65
12. Harzheim, D., Klose, H., Pinado, F.P. *et al.* Anxiety and depression disorders in patients with pulmonary arterial hypertension and chronic thromboembolic pulmonary hypertension. *Respir Res* **14,** 104 (2013). https://doi.org/10.1186/1465-9921-14-104
13. Mohammad Shoaib Hamrah, Mohammad Hassan Hamrah, Hideki Ishii, Susumu Suzuki, Mohammad Hussain Hamrah, Ahmad Edris Hamrah, Ahmad Elias Dahi, Kyosuke Takeshita, Maimaiti Yisireyili, Mohammad Hashem Hamrah, Akbar Fotouhi, Junichi Sakamoto, Toyoaki Murohara, "Anxiety and Depression among Hypertensive Outpatients in Afghanistan: A Cross-Sectional Study in Andkhoy City", International Journal of Hypertension, vol. 2018, Article ID 8560835, 8 pages, 2018. https://doi.org/10.1155/2018/8560835
14. Ashok VG, Ghosh SS. Prevalence of Depression among Hypertensive Patients Attending a Rural Health Centre in Kanyakumari. Natl J Community Med 2019;10(3):172-175
15. Roberts L, Davis GK and Homer CSE (2019) Depression, Anxiety, and Post-traumatic Stress Disorder Following a Hypertensive Disorder of Pregnancy: A Narrative Literature Review. *Front. Cardiovasc. Med.* 6:147. doi: 10.3389/fcvm.2019.001474
16. Yunping Li, Ranran Wang, Jingqun Tang, Chen Chen, Ling Tan, Zhongshi Wu, Fenglei Yu, Xiang Wang, "Progressive Muscle Relaxation Improves Anxiety and Depression of Pulmonary Arterial Hypertension Patients", Evidence-Based Complementary and Alternative Medicine, vol. 2015, Article ID 792895, 8 pages, 2015. https://doi.org/10.1155/2015/792895
17. Menezes P, Quayle J, Garcia Claro H, da Silva S, Brandt LR, Diez-Canseco F, Miranda JJ, Price LN, Mohr DC, Araya R. Use of a Mobile Phone App to Treat Depression Comorbid With Hypertension or Diabetes: A Pilot Study in Brazil and Peru. JMIR Ment Health 2019;6(4):e11698

18. Petersen, I., Bhana, A., Folb, N. *et al.* Collaborative care for the detection and management of depression among adults with hypertension in South Africa: study protocol for the PRIME-SA randomised controlled trial. *Trials* **19,** 192 (2018). https://doi.org/10.1186/s13063-018-2518-6
19. Gress TW, Nieto FJ, Shahar E, Wofford MR, Brancati FL. Hypertension and antihypertensive therapy as risk factors for type 2 diabetes mellitus. Atherosclerosis Risk in Communities Study.N Engl J Med. 2000; 342:905–912. doi: 10.1056/NEJM200003303421301.
20. Chen G, McAlister FA, Walker RL, Hemmelgarn BR, Campbell NR. Cardiovascular outcomes in framingham participants with diabetes: the importance of blood pressure.Hypertension. 2011; 57:891–897. doi: 10.1161/Hypertensionaha.110.162446.
21. Bassi, Daniela, Cabiddu, Ramona, Mendes, Renata G., Tossini, Natália, Arakelian, Vivian M., Caruso, Flávia C. R., Bonjorno Júnior, José C., Arena, Ross, & Borghi-Silva, Audrey. (2018). Effects of Coexistence Hypertension and Type II Diabetes on Heart Rate Variability and Cardiorespiratory Fitness. *Arquivos Brasileiros de Cardiologia*, *111*(1), 64-72.
22. Beryl Kemche, Brice Ulrich Saha Foudjo, Elie Fokou, "Risk Factors of Hypertension among Diabetic Patients from Yaoundé Central Hospital and Etoug-Ebe Baptist Health Centre, Cameroon", Journal of Diabetes Research, vol. 2020, Article ID 1853516, 8 pages, 2020. https://doi.org/10.1155/2020/1853516
23. Liu, CY., Zhang, W., Ji, LN. *et al.* Comparison between newly diagnosed hypertension in diabetes and newly diagnosed diabetes in hypertension. *Diabetol Metab Syndr* **11,** 69 (2019). https://doi.org/10.1186/s13098-019-0465-3
24. Ndejjo R, Masengere P, Nuwaha F et al. Hypertension and diabetes patients' perspective of challenges and their coping mechanisms in Mukono and Buikwe districts in Uganda – a qualitative study [version 1; peer review: 1 approved with reservations]. Open Research Europe 2021, 1:30 (https://doi.org/10.12688/openreseurope.13286.1)
25. Viswanathan Mohan, Yackoob K. Seedat, Rajendra Pradeepa, "The Rising Burden of Diabetes and Hypertension in Southeast Asian and African Regions: Need for Effective Strategies for Prevention and Control in Primary Health Care Settings", International Journal of Hypertension, vol. 2013, Article ID 409083, 14 pages, 2013. https://doi.org/10.1155/2013/409083.

Chapter 5

Hypertension and COVID-19: What does Research say?

Shivsanjeevi Sripathi

"Hope can be a powerful force. Maybe there's no actual magic in it, but when you know what you hope for most and hold it like a light within you, you can make things happen, almost like magic."

– Laini Taylor

CONTENTS

5.1 The Extent of Damage 220
5.2 Hypertension Sets the Stage 222
5.3 Perspectives of Hypertensive Patients 248
5.4 References 261

5.1 THE EXTENT OF DAMAGE

Severe Acute Respiratory Syndrome-Coronavirus-2 (SARS-CoV-2) as the causative agent of Coronavirus disease-2019 (COVID-19) has wreaked evident havoc across the globe with health and economic damage. Among the various risk factors of the current COVID-19 pandemic, a retrospective and prospective cohort study was reported as *"a first assessment of the information gathered on COVID-19 sequelae". The sample was inclusive of* the COVID-BioB study at San Raffaele University Hospital in Milan, Italy with COVID-19 detected in the patients from 7 April 2020 to 7 May 2020 by reverse-transcriptase polymerase chain reaction (RT-PCR) via nasopharyngeal and/or throat swabs. A slew of examinations was done ranging from breath shortness by the modified Medical Research Council (mMRC) scale for dyspnoea to the Mini Nutritional Assessment (MNA) screening to a visuo-analog scale (VAS) for self-rated health as depicted below in figure 1 (© 2020 De Lorenzo et al. *This is an open access article distributed under the terms of the Creative Commons Attribution License, which permits unrestricted use, distribution, and reproduction in any medium, provided the original author and source are credited*):

Key: mMRC, modified Medical Research Council. VAS, visuo-analog scale. MoCA, Montreal Cognitive Assessment. MNA, Mini Nutritional Assessment. IES-R, Impact of Event Scale-Revised. STAI-Y, State-Trait Anxiety Inventory form Y. WHIIRS, Women's Health Initiative Insomnia Rating Scale

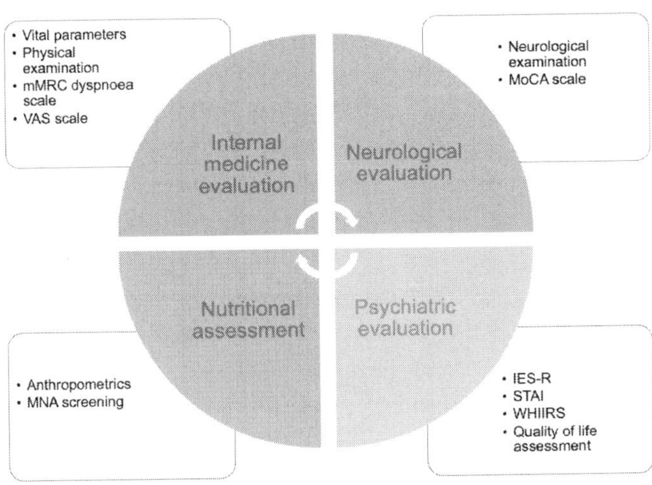

The general aspects of the patients (n=185) are represented below in figure 2 (© 2020 De Lorenzo et al. *This is an open access article distributed*

under the terms of the Creative Commons Attribution License, which permits unrestricted use, distribution, and reproduction in any medium, provided the original author and source are credited):

Variable	All Cohort	Discharged from ED	Hospitalised	P
No. of patients	185 (100)	59 (31.9)	126 (68.1)	
Age (years)	57 (48; 67)	50 (40.5; 57.7)	61 (51.2; 69)	<0.0001
Female gender	62 (33.5)	28 (47.5)	34 (27)	0.0098
Ethnicity				0.029
European (Eastern/Western)	168 (90.8)	49 (83.1)	119 (94.4)	
Hispanic (South America)	16 (8.6)	9 (15.3)	7 (5.6)	
African-American (Black)	1 (0.5)	1 (1.7)	0 (0)	
BMI (Kg/m^2)	27.2 (24.7; 30.5)	26.2 (24.1; 30.7)	27.8 (25.4; 30.5)	0.12
Comorbidities				
HTN	70 (37.8)	15 (25.4)	55 (43.7)	0.029
CAD	12 (6.5)	5 (8.5)	7 (5.6)	0.65
DM	21 (11.4)	4 (6.8)	17 (13.5)	0.28
COPD	2 (1.1)	0 (0)	2 (1.6)	0.84
CKD	3 (1.6)	0 (0)	3 (2.4)	0.58
Active cancer	3 (1.6)	1 (1.7)	2 (1.6)	1.00
Psychiatric disorder	40 (21.6)	15 (25.4)	25 (19.8)	0.77
At ED presentation				
PaO_2/FiO_2	314.8 (263.8; 357.6)	371.4 (333.8; 409.5)	296.4 (249.6; 330.7)	<0.0001
Body temperature (°C)	37.8 (37; 38.3)	37.5 (36.6; 38)	37.9 (37.1; 38.5)	0.0088
WBC (x10^9/L)	6 (4.8; 7.9)	5.5 (4.2; 6.6)	6.3 (4.9; 8.3)	0.017
NLR	3.5 (2; 6.3)	2.7 (1.9; 3.9)	4.4 (2.2; 7.3)	0.0018
AST (U/L)	39 (27; 55.2)	27 (24; 47)	44 (32; 58.5)	<0.0001
ALT (U/L)	35 (22.8; 55)	27 (21; 48)	38 (24; 58)	0.058
LDH (U/L)	330 (248; 409.5)	249 (208; 340)	362.5 (281.2; 426)	<0.0001
CRP (mg/dL)	46.7 (14.8; 100.8)	12.3 (4.9; 48.8)	67.2 (29; 120.2)	<0.0001
eGFR (mL/min)	85.5 (68.3; 97.2)	92.8 (76.2; 105.8)	79.8 (66.4; 92.4)	0.0001
LoS	9.5 (6; 15)	-	9.5 (6; 15)	-
Transfer to ICU	4 (2.2)	-	4 (3.2)	-
NIV	32 (17.3)	-	32 (25.4)	-

Dichotomous variables were expressed as count (percentage) and continuous variables as median (interquartile range).

Abbreviations: BMI, body mass index; ED, Emergency Department; PaO_2/FiO_2, arterial oxygen partial pressure/fractional inspired oxygen; WBC, white blood cell count; NLR, neutrophil to lymphocyte ratio; AST, aspartate aminotransferase; ALT, alanine aminotransferase; LDH, lactic dehydrogenase; CRP, C-reactive protein; eGFR, estimated glomerular filtration rate; HTN, arterial hypertension; CAD, coronary artery disease; DM, diabetes mellitus; COPD, chronic obstructive pulmonary disease; CKD, chronic kidney disease; LoS, length of stay; ICU, intensive care unit; NIV, non-invasive ventilation.

https://doi.org/10.1371/journal.pone.0239570.t001

An interesting observation was that the frequency of hypertension was more in patients who were admitted to the hospital vs. those who were sent off from the Emergency Department (ED). Further changes included malnourishment or its risk in 62.7% of the patients; breath shortness in 29.2%, cognitive impairment in 25% and mental health issues in 44.9% of the patients. The necessity of follow-up was required by uncontrolled hypertension, shortness of breath, malnutrition or the sudden cognitive impairment that was previously not noted. Figure 3 below is representative of the presence of hypertension as a change during follow-up with other conditions with the numbers representing the count of patients (© 2020 De Lorenzo et al. *This is an open access article distributed under the terms of the Creative Commons Attribution License, which permits unrestricted use, distribution, and reproduction in any medium, provided the original author and source are credited*):

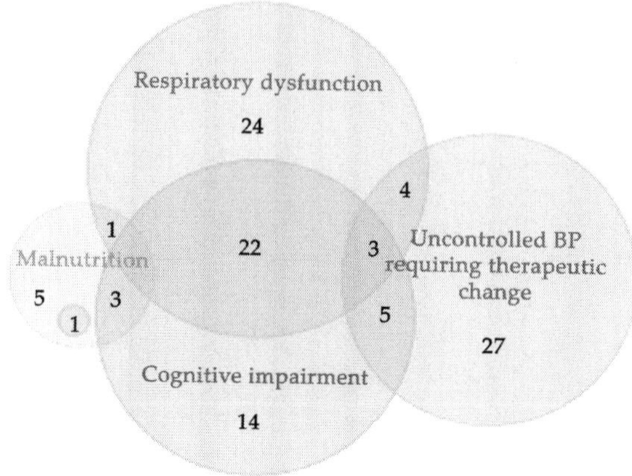

5.2 HYPERTENSION SETS THE STAGE

A comprehensive review of metabolic risk factors for COVID-19 was done following the Preferred Reporting Items for Systematic Reviews and Meta-Analyses (PRISMA) statement by Moazzami and team (2020). Analyses narrowed down on 13 articles exploring the extent of hypertension, obesity and diabetes with ten of these studies reporting the presence of hypertension: refer table 1 below (© 2020 Moazzami et al. *This is an open access article distributed under the terms of the Creative Commons Attribution License, which permits unrestricted use, distribution, and reproduction in any medium, provided the original author and source are credited*):

No	Author	Country, Year	Study Design	Study Scope	Sample Size	Age mean/ median (year)	Men proportion (n, %)	Measurement of Covid-19*	Obesity Prevalence (n, %)	Diabetes Prevalence (n, %)	Hypertension Prevalence (n, %)	Hyper lipidemia Prevalence (n, %)
1	Alattar R, et al. [18]	Qatar, 2020	Retrospective Cohort	Local	25	58(50–63)	23(92%)	1	-	12(48%)	-	-
2	Barrasa H, et al. [9]	Spain, 2020	Cohort	National	48	63 (12)	27 (56%)	5 & 1	23 (48%)	9 (19%)	21 (44%)	-
3	CDC COVID-19 Response Team [19]	USA, 2020	Cross-Sectional	National	7162	-	-	1	-	10.9%	-	-
4	Claire E Hastie, et al. [20]	England, Scotland and Wales, 2020	Cross-Sectional	Regional	449	(37–73)	256(59.02%)	1	158 (35.19%)	400 (89.09%)	-	-
5	Gentile S, et al. [5]	Italy, 2020	Cross-Sectional	National	1890	-	-	1	12.2%	31.9%	69.7%	-
6	GuanW J, et al. [21]	China, 2020	Retrospective	National	1590	48.9(16.3)	904(57.3%)	1	-	8.2%	16.9%	-
7	Guo W, et al. [22]	China, 2020	Retrospective	Local	174	59 (49–67)	76 (43.7%)	1,2	-	37(21.2%)	43 (24.7%)	-
8	Hong K S, et al. [23]	South Korea, 2020	Retrospective Cohort	Local	98	55.4(17.1)	38(38.8%)	1	-	9(9.2%)	30(30.6%)	-
9	Nikpouraghdam M, et al. [24]	Iran, 2020	Retrospective	Local	2964	55.50 (15.15)	1955(66%)	1	-	113(3.81%)	59(1.99%)	-
10	Sean Wei Xiang Ong, et al. [6]	Singapore, 2020	Retrospective Cohort	Local	91	(NM-60)	-	1	12.1%	18(19.8%)	30(33%)	-
11	Shi, Y, et al. [25]	China, 2020	Retrospective Cohort	Local	487	46 (19)	259 (53.2%)	2 & 3 & 4	-	29 (6.0%)	99 (20.3%)	-
12	Simonnet A, et al [26]	France, 2020	Retrospective Cohort	Local	124	60(51–70)	73%	1 & 3	47.6%	28 (23%)	60 (49%)	34 (28%)
13	Yang zhang, et al. [7]	China, 2020	Retrospective	Local	166	62.7(14.2)	85(51.2%)	3,4	-	21.1%	76(45.8%)	-

*measure type (PCR: 1, CT: 2, clinical: 3, Lab: 4, World Health Organization interim guidance by RT-PCR: 5)

https://doi.org/10.1371/journal.pone.0243600.t002

The pooled prevalence of hypertension was 32% while that for obesity was 29% and that of diabetes was 22% that led the authors to conclude that "*in other words, that hypertension is more prevalent than obesity and diabetes in patients with Covid-19 infection*" (Moazzami et al, 2020).

The analyses for the prevalence of hypertension in COVID-19 research articles are presented below: ***TOP***: Pooled prevalence (© 2020 Moazzami et al. *This is an open access article distributed under the terms of the Creative Commons Attribution License, which permits unrestricted use, distribution, and reproduction in any medium, provided the original author and source are credited*):

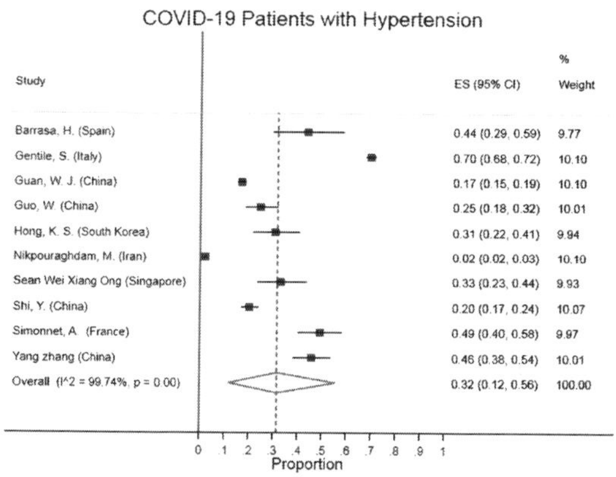

BOTTOM: Sensitivity by Forest plots (© 2020 Moazzami et al. *This is an open access article distributed under the terms of the Creative Commons Attribution License, which permits unrestricted use, distribution, and reproduction in any medium, provided the original author and source are credited*):

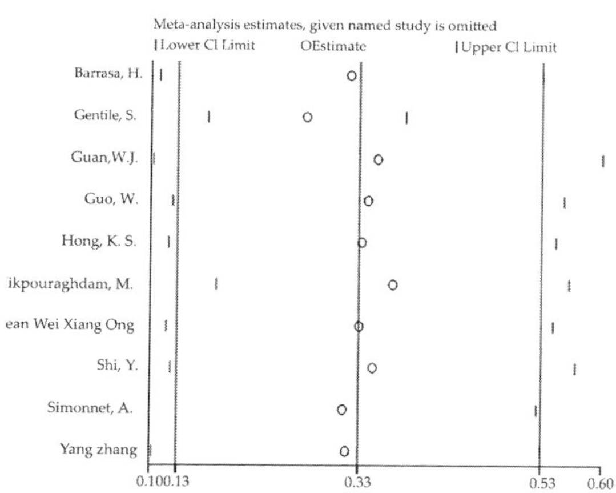

Another meta-analysis published in *Diabetology & Metabolic Syndrome* (2020) by de Almeida-Pititto was described as "*one of the largest meta-analysis, to our knowledge, associating cardiovascular disease, diabetes and hypertension with COVID-19 outcomes*". On the basis of the International Prospective Register of Systematic Reviews (PROSPERO: CRD42020184254), 40 studies were pinpointed that explored the outcome and association of hypertension and other comorbidities in the condition. The studies are represented below in table 2 (de Almeida-Pititto et al, 2020: *Open Access: This article is licensed under a Creative Commons Attribution 4.0 International License, which permits use, sharing, adaptation, distribution and reproduction in any medium or format, as long as you give appropriate credit to the original author(s) and the source, provide a link to the Creative Commons licence, and indicate if changes were made. The images or other third party material in this article are included in the article's Creative Commons licence, unless indicated otherwise in a credit line to the material. If material is not included in the article's Creative Commons licence and your intended use is not permitted by statutory regulation or exceeds the permitted use, you will need to obtain permission directly from the copyright holder. To view a copy of this licence, visit http://creativecommons.org/licenses/by/4.0/. The Creative Commons Public Domain Dedication waiver (http://creativecommons.org/publicdomain/zero/1.0/) applies to the data made available in this article, unless otherwise stated in a credit line to the data*):

Key: *DM, diabetes mellitus; HAS, hypertension; DCV, cardiovascular disease; ACEI/ARB, angiotensin converting enzyme inhibitors/angiotensin II receptor blockers*

Reference numbers	**First author**	**Type of study**	**Location**	**Sample size**	**Morbidities (HAS/DM/DCV/ACE OU BRA)**	**Outcome (severity and/or mortality)**
[14]	Li et al.	Case Series	CHINA	362	HAS/DM/DCV/ACEI/ARB uses	Severity and Mortality
[15]	Giamarellos-Bourboulis et al.	Case Series	GREECE	54	DM/DCV	Severity

[16]	Zhang et al.	Retrospective Cohort	CHINA	19	HAS/DM/DCV	Mortality
[17]	Zhou et al.	Retrospective Cohort	CHINA	191	HAS/DM/DCV	Mortality
[18]	Colombi et al.	Retrospective Cohort	ITALY	236	DM/DCV	Severity
[19]	Li et al.	Cross Sectional	CHINA	548	HAS/DM/DCV/ACEI/ARB use	Severity
[20]	Lei et al.	Retrospective Cohort	CHINA	34	DM/DCV	Severity
[21]	Deng et al.	Retrospective Cohort	CHINA	112	HAS/DM/DCV	Severity
[22]	Guo et al.	Case Series	CHINA	187	HAS/DM/DCV/ACEI/ARB use	Severity
[23]	Huang et al.	Retrospective Cohort	CHINA	41	HAS/DM/DCV	Severity
[24]	Zhang et al.	Case Series	CHINA	140	HAS/DM/DCV	Severity
[25]	Liu et al.	Retrospective Cohort	CHINA	78	HAS/DM/DCV	Severity
[26]	Wang et al.	Retrospective Cohort	CHINA	138	HAS/DM/DCV	Severity
[27]	Guan et al.	Retrospective Cohort	CHINA	1099	HAS/DM/DCV	Severity and Mortality
[28]	Wu et al.	Retrospective Cohort	CHINA	84	HAS/DM/DCV	Mortality
[29]	Zhang et al.	Retrospective Cohort	CHINA	221	HAS/DM/DCV	Severity
[30]	Chen et al.	Retrospective Cohort	CHINA	274	HAS/DM/DCV	Mortality
[31]	Deng et al.	Retrospective Cohort	CHINA	225	HAS/DM/DCV	Mortality
[32]	Wang et al.	Case Series	CHINA	69	HAS/DM/DCV	Severity
[33]	Yang et al.	Retrospective Cohort	CHINA	52	DM/DCV	Mortality
[34]	Guan et al.	Case Series	CHINA	1590	DM/DCV	Mortality

[35]	Zheng et al.	Retrospective Cohort	CHINA	161	HAS/DM/DCV	Severity
[36]	Fan et al.	Retrospective Cohort	CHINA	21	HAS/DM/DCV	Mortality
[37]	Yuan et al.	Case Series	CHINA	27	HAS/DM/DCV	Severity
[38]	Feng et al.	Retrospective Cohort	CHINA	476	HAS/DM/DCV	Severity
[39]	Mao et al.	Case Series	CHINA	214	HAS/DM/DCV	Severity
[40]	Wang et al.	Retrospective Cohort	CHINA	339	HAS/DM/DCV	Mortality
[41]	Sim-onet et al.	Retrospective Cohort	FRANCE	124	HAS/DM/DCV	Severity
[42]	Chen et al.	Cross Sec-tional	CHINA	150	HAS	Severity
[43]	Wu et al.	Retrospective Cohort	CHINA	201	HAS	Severity
[44]	Xiang et al.	Cross Sec-tional	CHINA	49	HAS	Severity
[45]	Wan et al.	Cross Sec-tional	CHINA	135	HAS/DM/DCV	Severity
[46]	Qin et al.	Retrospective Cohort	CHINA	452	HAS	Severity
[47]	Ruan et al.	Retrospective Cohort	CHINA	150	HAS	Mortality
[48]	Zhang et al.	Retrospective Cohort	CHINA	1128	ACEI/ARB use	Mortality
[49]	Xu et al.	Retrospective Cohort	CHINA	187	HAS	Mortality
[50]	Meng et al.	Case series	CHINA	42	ACEI/ARB use	Severity
[51]	Yang et al.	Retrospective Cohort	CHINA	125	ACEI/ARB use	Severity
[52]	Mancia et al.	Retrospective Cohort	ITALY	6272	ACEI/ARB use	Severity
[53]	Reyn-olds et al.	Retrospective Cohort	USA	2005	ACEI/ARB use	Severity

The analyses of these studies inclusive of 18.012 COVID patients (1315 with hypertension) led to the researchers ***identifying a moderate positive association between hypertension and the severity and mortality***

of COVID-19: while diabetes displayed the same trend, the association was strong for previous cardiovascular disease. The meta-analyses is depicted below in figure 4 (de Almeida-Pititto et al, 2020: *Open Access: This article is licensed under a Creative Commons Attribution 4.0 International License, which permits use, sharing, adaptation, distribution and reproduction in any medium or format, as long as you give appropriate credit to the original author(s) and the source, provide a link to the Creative Commons licence, and indicate if changes were made. The images or other third party material in this article are included in the article's Creative Commons licence, unless indicated otherwise in a credit line to the material. If material is not included in the article's Creative Commons licence and your intended use is not permitted by statutory regulation or exceeds the permitted use, you will need to obtain permission directly from the copyright holder. To view a copy of this licence, visit http://creativecommons.org/licenses/by/4.0/. The Creative Commons Public Domain Dedication waiver (http://creativecommons.org/publicdomain/zero/1.0/) applies to the data made available in this article, unless otherwise stated in a credit line to the data*):

(A)

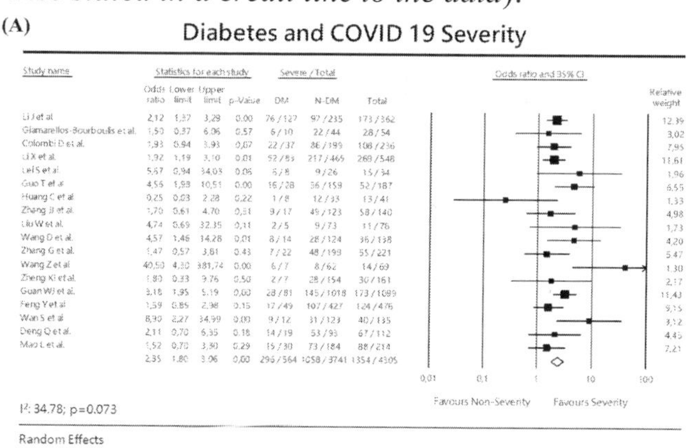

(B)

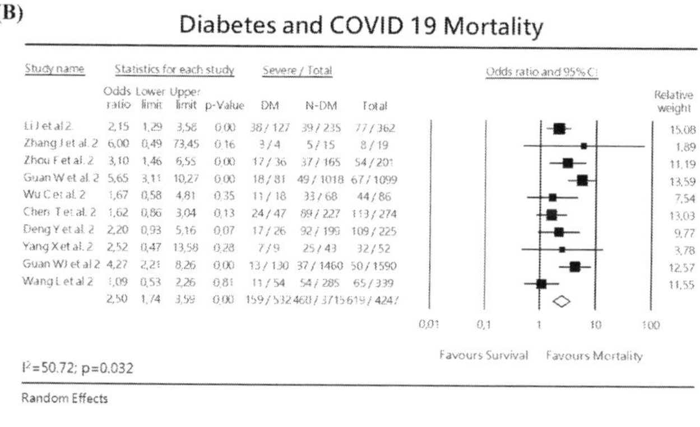

(C)

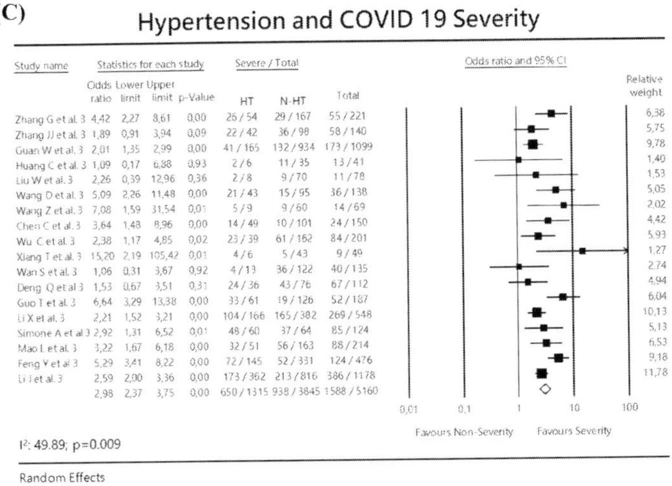
Hypertension and COVID 19 Severity
Study name
Statistics for each study
Severe / Total
Odds ratio and 95% CI
Odds ratio Lower limit Upper limit p-Value
HT N-HT Total
Relative weight
Zhang G et al. 3 4,42 2,27 8,61 0,00 26 / 54 29 / 167 55 / 221 6,38
Zhang JJ et al. 3 1,89 0,91 3,94 0,09 22 / 42 36 / 98 58 / 140 5,75
Guan W et al. 3 2,01 1,35 2,99 0,00 41 / 165 132 / 934 173 / 1099 9,78
Huang C et al. 3 1,09 0,17 6,88 0,93 2 / 6 11 / 35 13 / 41 1,40
Liu W et al. 3 2,26 0,39 12,96 0,36 2 / 8 9 / 70 11 / 78 1,53
Wang D et al. 3 5,09 2,26 11,48 0,00 21 / 43 15 / 95 36 / 138 5,05
Wang Z et al. 3 7,08 1,59 31,54 0,01 5 / 9 9 / 60 14 / 69 2,02
Chen C et al. 3 3,64 1,48 8,96 0,00 14 / 49 10 / 101 24 / 150 4,42
Wu C et al. 3 2,38 1,17 4,85 0,02 23 / 39 61 / 162 84 / 201 5,93
Xiang T et al. 3 15,20 2,19 105,42 0,01 4 / 6 5 / 43 9 / 49 1,27
Wan S et al. 3 1,06 0,31 3,67 0,92 4 / 13 36 / 122 40 / 135 2,74
Deng Q et al 3 1,53 0,67 3,51 0,31 24 / 36 43 / 76 67 / 112 4,94
Guo T et al. 3 6,64 3,29 13,38 0,00 33 / 61 19 / 126 52 / 187 6,04
Li X et al. 3 2,21 1,52 3,21 0,00 104 / 166 165 / 382 269 / 548 10,13
Simone A et al 3 2,92 1,31 6,52 0,01 48 / 60 37 / 64 85 / 124 5,13
Mao L et al. 3 3,22 1,67 6,18 0,00 32 / 51 56 / 163 88 / 214 6,53
Feng Y et al 3 5,29 3,41 8,22 0,00 72 / 145 52 / 331 124 / 476 9,18
Li J et al. 3 2,59 2,00 3,36 0,00 173 / 362 213 / 816 386 / 1178 11,78
2,98 2,37 3,75 0,00 650 / 1315 938 / 3845 1588 / 5160
0,01 0,1 1 10 100
Favours Non-Severity Favours Severity
I²: 49.89; p=0.009
Random Effects

(D)

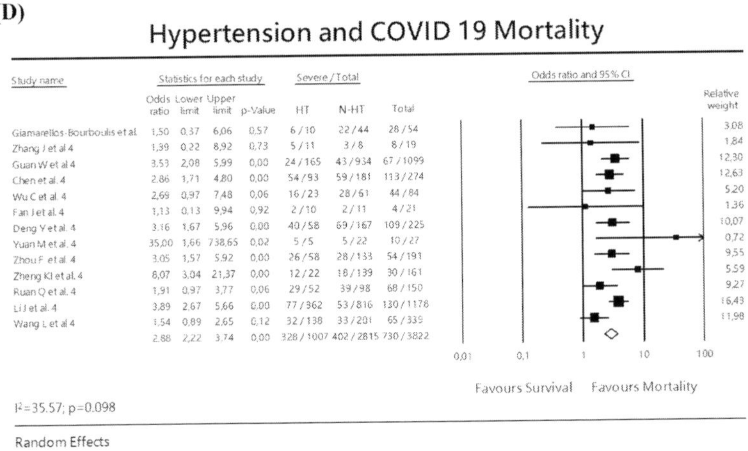
Hypertension and COVID 19 Mortality
Study name
Statistics for each study
Severe / Total
Odds ratio and 95% CI
Odds ratio Lower limit Upper limit p-Value
HT N-HT Total
Relative weight
Giamarellos-Bourboulis et al. 1,50 0,37 6,06 0,57 6 / 10 22 / 44 28 / 54 3,08
Zhang J et al 4 1,39 0,22 8,92 0,73 5 / 11 3 / 8 8 / 19 1,84
Guan W et al 4 3,53 2,08 5,99 0,00 24 / 165 43 / 934 67 / 1099 12,30
Chen et al. 4 2,86 1,71 4,80 0,00 54 / 93 59 / 181 113 / 274 12,63
Wu C et al. 4 2,69 0,97 7,48 0,06 16 / 23 28 / 61 44 / 84 5,20
Fan J et al. 4 1,13 0,13 9,94 0,92 2 / 10 2 / 11 4 / 21 1,36
Deng Y et al. 4 3,16 1,67 5,96 0,00 40 / 58 69 / 167 109 / 225 10,07
Yuan M et al. 4 35,00 1,66 738,65 0,02 5 / 5 5 / 22 10 / 27 0,72
Zhou F et al. 4 3,05 1,57 5,92 0,00 26 / 58 28 / 133 54 / 191 9,55
Zheng KI et al. 4 8,07 3,04 21,37 0,00 12 / 22 18 / 139 30 / 161 5,59
Ruan Q et al. 4 1,91 0,97 3,77 0,06 29 / 52 39 / 98 68 / 150 9,27
Li J et al. 4 3,89 2,67 5,66 0,00 77 / 362 53 / 816 130 / 1178 16,43
Wang L et al 4 1,54 0,89 2,65 0,12 32 / 138 33 / 201 65 / 339 11,98
2,88 2,22 3,74 0,00 328 / 1007 402 / 2815 730 / 3822
0,01 0,1 1 10 100
Favours Survival Favours Mortality
I²=35.57; p=0.098
Random Effects

(E)

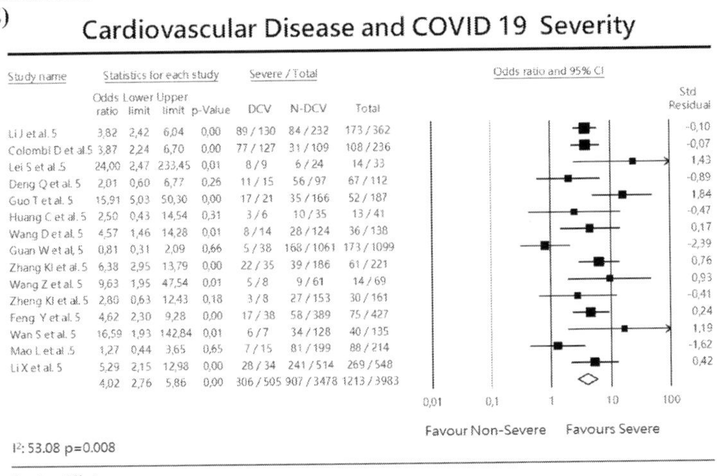
Cardiovascular Disease and COVID 19 Severity
Study name
Statistics for each study
Severe / Total
Odds ratio and 95% CI
Odds ratio Lower limit Upper limit p-Value
DCV N-DCV Total
Std Residual
Li J et al. 5 3,82 2,42 6,04 0,00 89 / 130 84 / 232 173 / 362 -0,10
Colombi D et al. 5 3,87 2,24 6,70 0,00 77 / 127 31 / 109 108 / 236 -0,07
Lei S et al. 5 24,00 2,47 233,45 0,01 8 / 9 6 / 24 14 / 33 1,43
Deng Q et al. 5 2,01 0,60 6,77 0,26 11 / 15 56 / 97 67 / 112 -0,89
Guo T et al. 5 15,91 5,03 50,30 0,00 17 / 21 35 / 166 52 / 187 1,84
Huang C et al. 5 2,50 0,43 14,54 0,31 3 / 6 10 / 35 13 / 41 -0,47
Wang D et al. 5 4,57 1,46 14,28 0,01 8 / 14 28 / 124 36 / 138 0,17
Guan W et al. 5 0,81 0,31 2,09 0,66 5 / 38 168 / 1061 173 / 1099 -2,39
Zhang KI et al. 5 6,38 2,95 13,79 0,00 22 / 35 39 / 186 61 / 221 0,76
Wang Z et al. 5 9,63 1,95 47,54 0,01 5 / 8 9 / 61 14 / 69 0,93
Zheng KI et al. 5 2,80 0,63 12,43 0,18 3 / 8 27 / 153 30 / 161 -0,41
Feng Y et al. 5 4,62 2,30 9,28 0,00 17 / 38 58 / 389 75 / 427 0,24
Wan S et al. 5 16,59 1,93 142,84 0,01 6 / 7 34 / 128 40 / 135 1,19
Mao L et al. 5 1,27 0,44 3,65 0,65 7 / 15 81 / 199 88 / 214 -1,62
Li X et al. 5 5,29 2,15 12,98 0,00 28 / 34 241 / 514 269 / 548 0,42
4,02 2,76 5,86 0,00 306 / 505 907 / 3478 1213 / 3983
0,01 0,1 1 10 100
Favour Non-Severe Favours Severe
I²: 53.08 p=0.008
Random Effects

(F)

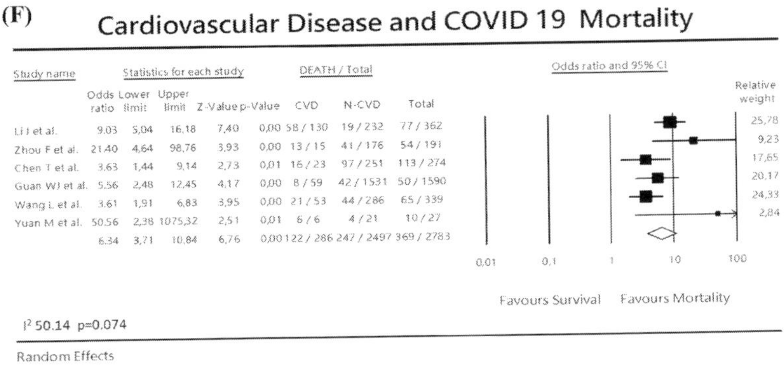

(G)

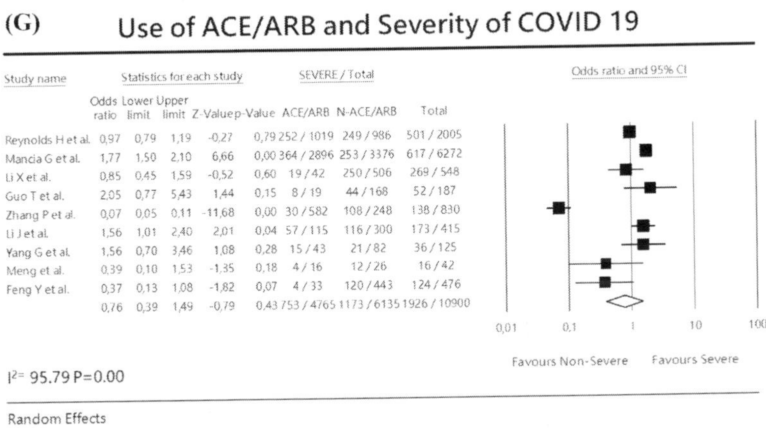

Another meta-analysis by Emami Amir et al (2020) in the *Archives of academic emergency medicine* scrutinized the prevalence of underlying diseases in ten studies with 3,403 hospitalized patients.

Table 3 below presents the studies looking at underlying diseases in COVID-19 patients (Emami Amir et al, 2020: *This is an Open Access article distributed under the terms of the Creative Commons Attribution License, (http://creativecommons.org/licenses/by/3.0/) which permits unrestricted use, distribution, and reproduction in any medium, provided the original work is properly cited*):

First Author	n	Sex (M/F)	Age	CKD	HTN	DM	Mal	Smoking	COPD	CVD
Chaolin Huang, et al. (34)	41	30/11	Range: 41-58	3	6	8	1	3	1	6
Nanshan Chen, et al. (2)	99	67/32	Mean: 55.5 ±13.1	3			1	2		40

Dawei Wang, et al. (35)	138	78/63	Median: 56 (42-68)	4	43	14	10		4	20
Jie.Li, et al. (36)	17	9/8	Range: 22-65	8	1			3		
Wei-Jie Guan, et al. (37)	1099	640/459	Range: 35-58	1	164	81	10	158	12	27
Xiao-Wei Xu, et al. (38)	62	35/27	Median: 41 (32-52)		5	1			1	1
Wenhua Liang,et al. (14)	1590	-	-				18	111		
Jin-jin Zhang, et al. (16)	140	71/69	Median: 57 (25-87)	2	42	17		9	2	7
Jian Wu, et al. (39)	80	39/41	Mean: 46.1 ± 15.4	1			1			25
Kui L, et al. (40)	137	61/76	Median: 57(20-83)		13	14	2			10

Figure 5 below shows the occurrence of hypertension emerging as the most prevalent disease in COVID-19 patients with 16% of hospitalized patients showing the condition. Other conditions identified were cardiovascular disease, diabetes, kidney disease, smoking, and COPDs ((chronic obstructive pulmonary diseases) (Emami Amir et al, 2020: *This is an Open Access article distributed under the terms of the Creative Commons Attribution License, (http://creativecommons.org/licenses/by/3.0/) which permits unrestricted use, distribution, and reproduction in any medium, provided the original work is properly cited*):

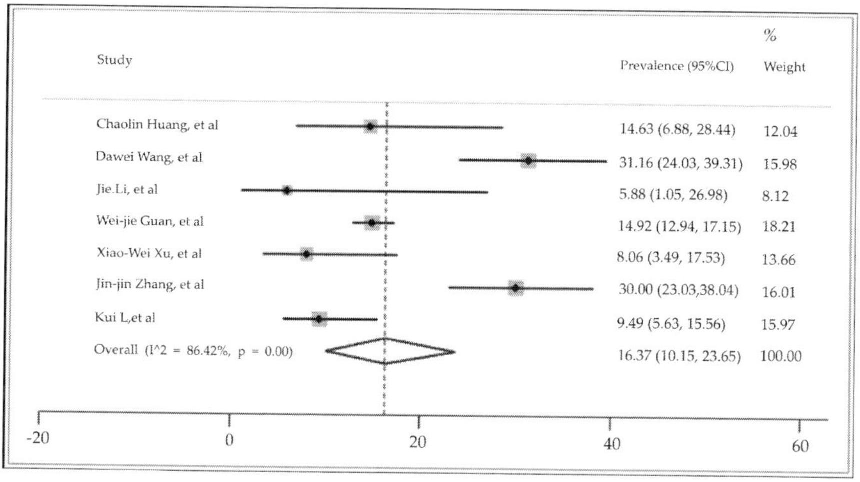

2021-published research in the *American journal of infection control* assayed for the trends in comorbidities and fatalities in COVID-19 patients between January 2020 and April 2020. 57.7% of the 22,753 COVID-19 positive patients displayed comorbidities with the prevalence of hypertension at 27.4%. The researchers reported that the common conditions were hypertension with diabetes and CVD (cardiovascular disease)/COPD (chronic obstructive pulmonary disease) at the following positions (Bajgain et al, 2021).

A retrospective analysis of the data of COVID-19 patients at the Department of Infectious Diseases, Renmin Hospital of Wuhan University, China was published in 2021 by Chen and team. From 366 patients who were hospitalized, exclusion resulted in 186 subject details that were assessed for the ***COVID-19 severity*** with that of ***hypertension*** (≥ 140/90 mmHg) and cardiac troponin I (cTnI: an indicator of cardiac injury at >40 pg/mL).

The system adopted is represented below in figure 6 (Copyright © 2021 Chen et al, *This is an open access article distributed under the terms of the Creative Commons Attribution License, which permits unrestricted use, distribution, and reproduction in any medium, provided the original author and source are credited*):

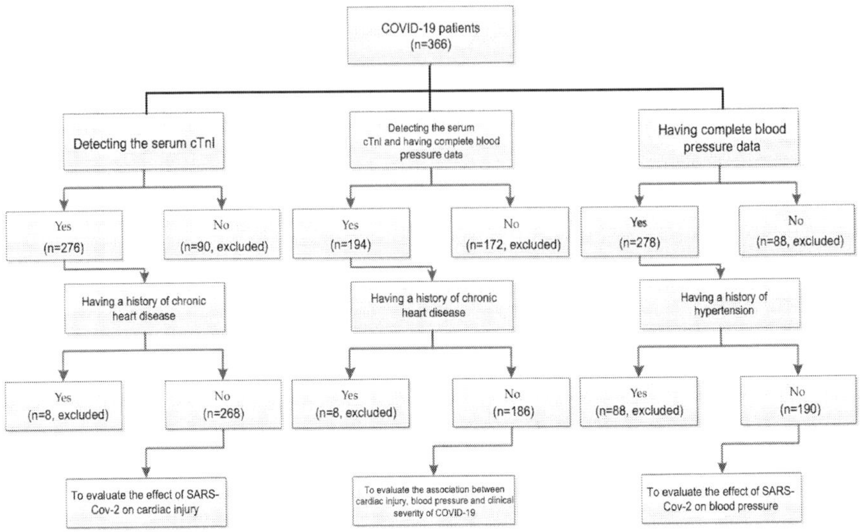

8.42% of 190 patients (16 total=10 ladies and 6 men) displayed a rise in blood pressure while in hospital. Most patients demonstrated a rise in the systolic pressure. Interestingly the levels of angiotensin II (Ang II: of the renin-angiotensin system) were higher in the patients vs. healthy controls. The trend in the blood pressure in COVID-19 patients as analyzed by the researchers is depicted below (Copyright © 2021 Chen et al, *This is an open access article distributed under the terms of the Creative Commons Attribution License, which permits unrestricted use, distribution, and reproduction in any medium, provided the original author and source are credited*):

A. Variations in the systolic blood pressure and cTnI; B. Changes in the systolic blood pressure, cTnI, white blood cells, and chest computed tomography in a patient: follow-up was 4 weeks post discharge

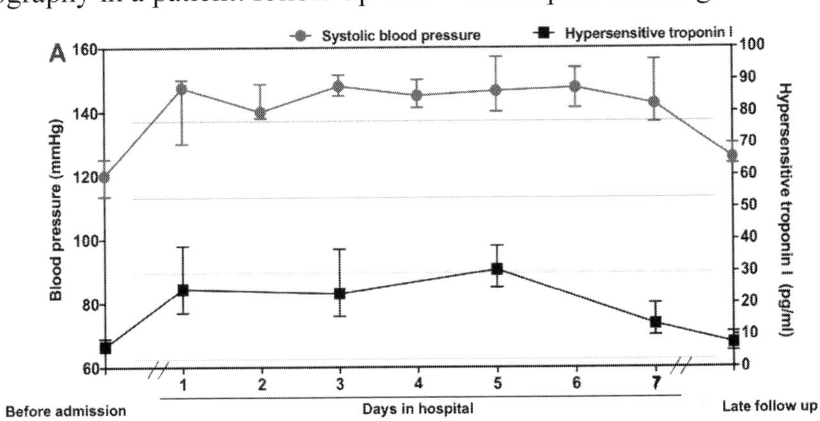

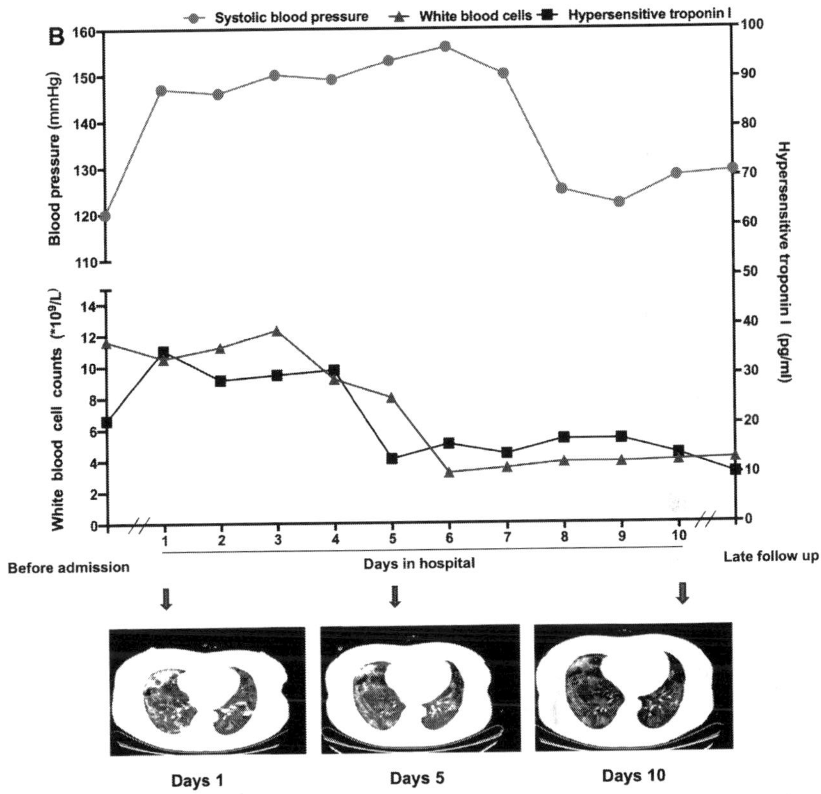

Table 4 below documents ***TOP***: the lack of conspicuous differences between the demographics at baseline in COVID-19 patients (Copyright © 2021 Chen et al, *This is an open access article distributed under the terms of the Creative Commons Attribution License, which permits unrestricted use, distribution, and reproduction in any medium, provided the original author and source are credited*):

Key: § Two-Sample T-test, £ Pearson's chi-square test, # continuous *correction Chi-square test*

	Total	**Blood pressure**	
	(n = 190)	**Normal (n = 174)**	**Elevated (n = 16)**
Age (y/o), Median (IQR)	54 (39–63)	53 (39–63)	60 (42–70)§
Gender (n, %)			
Male	78 (41.05)	72 (41.38)	6 (37.50)£

Female	112 (58.95)	102 (58.62)	10 (62.50)[£]
Clinical categories (n, %)			
Non-severe	116 (61.05)	108 (62.07)	8 (50.00)[£]
Severe	74 (38.95)	66 (37.93)	8 (50.00)[£]
Symptoms (n, %)			
Fever	156 (82.11)	142 (81.61)	14 (87.50)[#]
Cough	132 (69.47)	120 (68.97)	12 (75.00)[#]
Dyspnea	70 (36.84)	64 (36.78)	6 (37.50)[£]
Chest stuffiness	68 (35.79)	62 (35.63)	6 (37.50)[£]
Fatigue	96 (50.53)	88 (50.57)	8 (50.00)[£]
Muscle soreness	36 (18.95)	34 (19.54)	2 (12.50)[#]
Comorbidities (n, %)			
Diabetes	28 (14.74)	26 (14.94)	2 (12.50)[#]
Chronic lung diseases	6 (3.16)	6 (3.45)	0 (0)
Gastrointestinal diseases	4 (2.11)	4 (2.30)	0 (0)
Thyroid disease	2 (1.05)	2 (1.15)	0 (0)
Prostate disease	2 (1.05)	2 (1.15)	0 (0)

BOTTOM: Changes in cTnI and procalcitonin in COVID-19 patients with hypertension by the Mann-Whitney test ; $*P < 0.05$ is considered statistically significant (Copyright © 2021 Chen et al, *This is an open access article distributed under the terms of the Creative Commons Attribution License, which permits unrestricted use, distribution, and reproduction in any medium, provided the original author and source are credited*):

Laboratory findings	**Normal blood pressure**	**Elevated blood pressure**	**Normal range**
	(median, IQR)	**(median, IQR)**	
cTnI (pg/mL)	3.86 (2.49–5.15)	22.00 (18.20–30.00)*	0–40.00
White blood cells ($\times 10^9$cells/L)	5.24 (3.87–7.00)	4.86 (3.96–6.60)	3.50–9.50
Neutrophils ($\times 10^9$cells/L)	3.14 (2.48–4.96)	3.64 (3.06–4.94)	1.80–6.30
Lymphocytes ($\times 10^9$cells/L)	1.13 (0.76–1.56)	0.93 (0.57–1.23)*	1.10–3.20

Monocytes ($\times 10^9$cells/L)	0.44 (0.33–0.56)	0.55 (0.25–0.61)	0.10–0.60
Hemoglobin (g/L)	128 (118–136)	118(107–137)	115–150
Procalcitonin (pg/mL)	49 (28–73)	82 (53–430)*	0–100
C-reactive protein (mg/L)	20.00 (2.40–44.50)	7.80 (3.51–33.20)	0–10.00
Lactic dehydrogenase (U/L)	259 (206–313)	259 (208–289)	120–250

Table 5 below documents the changes in the ***TOP***: rennin-angiotensin system in COVID-19 patients vs. controls : § Two-Sample T-test, † Mann Whitney U test/Wilcoxon Sum Rank test (Copyright © 2021 Chen et al, *This is an open access article distributed under the terms of the Creative Commons Attribution License, which permits unrestricted use, distribution, and reproduction in any medium, provided the original author and source are credited*):

Laboratory findings	**Healthy control**	**Normal blood pressure**	**Elevated blood pressure**
	(median, IQR)	**(median, IQR)**	**(median, IQR)**
Adrenocorticotrophic hormone (pg/mL)	31.92 (21.6–39.67)	27.55 (21.12–39.56)§	33.03 (19.42–40.89)§
Renin (pg/mL)	5.31 (3.21–8.75)	6.35 (3.34–7.83)†	5.86 (4.41–6.76)†
Angiotensin II (pg/mL)	87.90 (48.23–107.39)	137.12. (123.63–161.67)*†	169.25 (142.17–186.98)*†
Aldosterone (pg/mL)	150.05 (129.32–164.32)	159.62 (119.16–169.30)†	141.54 (118.63–154.83)†

BOTTOM: rennin-angiotensin system in high BP vs. normal BP patients as assessed by the t-test (Copyright © 2021 Chen et al, *This is an open access article distributed under the terms of the Creative Commons Attribution License, which permits unrestricted use, distribution, and reproduction in any medium, provided the original author and source are credited*):

Laboratory findings	Total (median, IQR)	Normal blood pressure (median, IQR)	Elevated blood pressure (median, IQR)
Adrenocorticotrophic hormone (pg/mL)	30.56 (20.99–40.54)	27.55 (21.12–39.56)	33.03 (19.42–40.89)
Renin (pg/mL)	6.01 (4.15–7.46)	6.35 (3.34–7.83)	5.86 (4.41–6.76)
Angiotensin α (pg/mL)	153.67 (129.17–175.98)	137.12. (123.63–161.67)	169.25 (142.17–186.98)*
Aldosterone (pg/mL)	142.11 (119.16–163.28)	159.62 (119.16–169.30)	141.54 (118.63–154.83)

Analyses of severe vs. non-severe COVID-19 cases unveiled higher age, male gender, elevated levels of cTnI, white blood cells, neutrophils, procalcitonin, C-reactive protein, and lactic dehydrogenase in severe cases. The association of age, sex, cTnI, white blood cells, neutrophils, lymphocytes, C-reactive protein, lactic dehydrogenase, and history of hypertension and diabetes with severe COVID-19 was identified with subsequent multivariate analysis showing age, cTnI and history of hypertension and diabetes as independent predictors.

Table 6 below documents the results of the researchers of ***TOP***: severe vs. non-severe COVID-19 features (Copyright © 2021 Chen et al, *This is an open access article distributed under the terms of the Creative Commons Attribution License, which permits unrestricted use, distribution, and reproduction in any medium, provided the original author and source are credited*):

Key: § Two-Sample T-test, † Mann Whitney U test/Wilcoxon Sum Rank test, £ Pearson's chi-square test, # continuous correction Chi-square test

Clinical Severity		**Normal range**	
Non-severe	**Severe**		
(n = 111, 59.68%)	**(n = 75, 40.32%)**		
Age (y/o), Median (IQR)	42 (33–51)	66 (57–76)*§	-
Gender (n, %)			
Male	45 (40.54)	45 (60.00)*£	-
Female	66 (59.46)	30 (40.00)*£	-
Clinical categories (n, %)			
Elevated blood pressure	6 (5.41)	10 (13.33)£	-
Hypertension	15 (13.51)	20 (26.67)*£	-
Diabetes	7 (6.31)	21 (28.00)*£	-
Chronic lung diseases	7 (6.31)	10 (13.33)£	-
Chronic kidney diseases	2 (1.8)	2 (2.67)#	-
Gastrointestinal diseases	2 (1.8)	1 (1.33)#	-
Malignant tumor	2 (1.8)	0 (0)	-
Laboratory findings			
cTnI (pg/mL)	5.68 (4.62–6.45)	7.00 (5.78–27.00)*†	0–40.00
White blood cells ($\times 10^9$cells/L)	5.07 (3.65–6.00)	5.98 (4.60–10.00)*†	3.50–9.50
Neutrophils ($\times 10^9$cells/L)	2.67 (2.15–3.92)	3.85 (3.02–8.27)*†	1.80–6.30
Lymphocytes ($\times 10^9$cells/L)	1.27 (0.96–1.73)	0.78 (0.48–1.29)*†	1.10–3.20
Monocytes ($\times 10^9$cells/L)	0.42 (0.34–0.54)	0.44 (0.29–0.69)†	0.10–0.60
Procalcitonin (pg/mL)	36.00 (23.00–57.00)	67.50 (32.00–288.00)*†	0–100.00
C-reactive protein (mg/L)	6.20 (0.50–28.00)	48.55 (7.40–81.50)*†	0–10.00
Lactic dehydrogenase (U/L)	233.50 (187.50–292.00)	316.00 (235.00–454.00)*†	120–250

BOTTOM: Analyses of predictors of severe COVID-19 (Copyright © 2021 Chen et al, *This is an open access article distributed under the terms of the Creative Commons Attribution License, which permits unrestricted use, distribution, and reproduction in any medium, provided the original author and source are credited*):

	Odds ratio	**95% CI**	**P-value**
Univariate analysis			
Age (years)	1.12	1.09–1.15	<0.001*
Male (%)	2.20	1.21–4.00	0.010*
cTnI (pg/mL)	1.13	1.05 - 1.22	0.002*
White blood cells ($\times 10^9$cells/L)	1.39	1.13 - 1.70	0.002*
Neutrophils ($\times 10^9$cells/L)	1.50	1.19–1.90	0.001*
Lymphocytes ($\times 10^9$cells/L)	0.23	0.10–0.54	0.001*
Procalcitonin (pg/mL)	1.01	1.00–1.01	0.069
C-reactive protein (mg/L)	1.02	1.01–1.04	0.002*
Lactic dehydrogenase (U/L)	1.01	1.00–1.01	0.001*
Hypertension (%)	2.489	1.185–5.226	0.016*
Diabetes (%)	5.78	2.31–14.45	<0.001*
Multivariate analysis			
Age (years)	1.11	1.07–1.16	<0.001*
Male (%)	1.38	0.57–3.37	0.479
CTnI (pg/mL)	1.08	1.01–1.15	0.018*
Hypertension (%)	7.19	2.55–20.31	<0.001*
Diabetes (%)	4.28	1.41–12.97	0.010*

A retrospective analysis of 274 COVID-19 survivors from the Lagos State COVID-19 outpatient clinic seen between April and June 2020 by Bodunrin Osikomaiya and team (2021) in *BMC Infectious Diseases* to scan for **persistent COVID-19 symptoms** post-discharge. No co-morbidity was reported in 78.5% of the patients (age=10 to 83 years). The remaining showed hypertension and diabetes at 72.9% and 15.3%, respectively as co-morbidities. Regression revealed a lack of association of hypertension/ diabetes history among other factors in persistent COVID-19 symptoms (described as headache in 12.8%; chest pain and insomnia in 9.8% of the patients).

Table 7 below depicts the ***TOP***: persistent symptoms in COVID-19 survivors as assessed at the Lagos State clinic (Bodunrin Osikomaiya et al, 2021: *Open Access: This article is licensed under a Creative Commons Attribution 4.0 International License, which permits use, sharing, adaptation, distribution and reproduction in any medium or format, as long as you give appropriate credit to the original author(s) and the source, provide a link to the Creative Commons licence, and indicate if changes were made. The images or other third party material in this article are included in the article's Creative Commons licence, unless indicated otherwise in a credit line to the material. If material is not included in the article's Creative Commons licence and your intended use is not permitted by statutory regulation or exceeds the permitted use, you will need to obtain permission directly from the copyright holder. To view a copy of this licence, visit http:// creativecommons.org/licenses/by/4.0/. The Creative Commons Public Domain Dedication waiver (http://creativecommons.org/publicdomain/ zero/1.0/) applies to the data made available in this article, unless otherwise stated in a credit line to the data*):

System	**Symptom**	**N=274 (%)**
General	Fever	17 (6.2)
	Easy fatigability	35 (12.8)
	Weight loss	17 (6.2)
	Malaise	15 (5.5)
	Myalgia	24 (8.8)
	Excessive night sweat	10 (3.7)
Respiratory	Cough	25 (9.2)
	Dyspnea	26 (9.5)
	Chest pain	27 (9.8)
Otolaryngology (ORL)	Sore throat	11 (4.0)
	Dysgeusia	7 (2.6)
	Anosmia	5 (1.8)

Gastrointestinal	Loss of appetite	24 (8.8)
	Vomiting	2 (0.73)
	Nausea	6 (2.2)
	Diarrhoea	11 (4.0)
	Abdominal discomfort	17 (6.2)
Cardiovascular	Palpitations	20 (7.4)
Neurologic	Headache	35 (12.8)
	Insomnia	27 (9.8)
	Attention or memory deficit	14 (5.2)
	Dizziness	8 (2.9)
	Visual blurring	6 (2.2)
	Tremulousness	2 (0.7)
	Vertigo	3 (1.1)

BOTTOM: hypertension in post-COVID-19 survivors (Bodunrin Osikomaiya et al, 2021: *Open Access: This article is licensed under a Creative Commons Attribution 4.0 International License, which permits use, sharing, adaptation, distribution and reproduction in any medium or format, as long as you give appropriate credit to the original author(s) and the source, provide a link to the Creative Commons licence, and indicate if changes were made. The images or other third party material in this article are included in the article's Creative Commons licence, unless indicated otherwise in a credit line to the material. If material is not included in the article's Creative Commons licence and your intended use is not permitted by statutory regulation or exceeds the permitted use, you will need to obtain permission directly from the copyright holder. To view a copy of this licence, visit http://creativecommons.org/licenses/by/4.0/. The Creative Commons Public Domain Dedication waiver (http://creativecommons.org/publicdomain/zero/1.0/) applies to the data made available in this article, unless otherwise stated in a credit line to the data*):

Key: [a] both *DM and HTN.* [b] Other diseases like sickle cell disease, Human Immunodeficiency Virus, chronic heart disease, glaucoma, and a psychiatric issue

Patient Characteristics	**N=274 (%)**
Age [in years]	
≤35	92 (33.6)
>35–49	105 (38.3)
50 and above	77 (28.1)
Sex	
Male	181 (66.1)
Female	93 (33.9)
Comorbidity	
Present	59 (21.5)
Absent	215 (78.5)
[a]**Comorbidity**	*n*=59
Diabetes (DM)	9 (15.3)
Hypertension (HTN)	43 (72.9)
Asthma	2 (3.4)
Peptic Ulcer Disease	4 (6.8)
Rheumatoid Arthritis	3 (5.1)
[b]Other	5 (8.5)
Initial COVID-19 severity	
Asymptomatic	21 (7.7)
Mild	139 (50.7)
Moderate	107 (39.0)
Severe	7 (2.6)
Number of comorbid conditions	
No comorbid condition	215 (78.5)
1 comorbid condition	52 (19.0)
2 or more comorbid conditions	7 (2.5)
Persistent symptoms	
Present	112 (40.9)
Absent	162 (59.1)
Number of Persistent symptoms	
Absent	162 (59.1)
1 persistent symptom	31 (11.3)
2 persistent symptoms	27 (9.9)
3 or more persistent symptoms	54 (19.7)
Days after discharge, median [mean; SD]	15 days [mean: 15.6; SD±6.6]

Sumaira Mubarik and team (2021) published an analysis of the trend of ***hypertension*** in 1,833 COVID-19 patients who lost their lives across 200 medical centers (community centers or hospitals) during the December 2019-February 2020 phase of the early outbreak in Wuhan, China. 40.5% of these patients showed hypertension with these patients showing a higher prevalence of retirees, unemployed and those doing housework. Critical COVID-19 illness was observed in more than 25% of those with hypertension vs. 1/8th with no hypertension. The hypertensive group also demonstrated a higher number of mortalities vs. those without hypertension in the bulk period of mortality (first two weeks of February) that too in 15 days in the former group vs. 17 days in the latter from onset to death.

Table 8 below is demonstrative of ***TOP***: the features of the deceased patients at Wuhan during the early stages of the COVID-19 spread (© 2021 Mubarik, Liu, Eshak, Liu, Liu, Wang, Shi, Wen, Bai, Yu and Cao. *This is an open-access article distributed under the terms of the **Creative Commons Attribution License (CC BY)**. The use, distribution or reproduction in other forums is permitted, provided the original author(s) and the copyright owner(s) are credited and that the original publication in this journal is cited, in accordance with accepted academic practice. No use, distribution or reproduction is permitted which does not comply with these terms*):

Parameters	**Unmatched (before propensity score-matching)**					**Matched (after propensity score-matching)**				
	All patients (*n* = 1,833)	**Patients without hypertension (*n* = 1,091)**	**Patients with hypertension (*n* = 742)**	***p*-value***	**Standardized**** difference**	**All patients (*n* = 1,252)**	**Patients without hypertension (*n* = 783)**	**Patients with hypertension (*n* = 469)**	***p*-value***	**Standardized**** difference**
Age, years M(IQR)	70 (63,79)	72 (64,78)	73 (66,80)	0.56	0.10	71 (64,81)	71 (65,81)	71 (63,80)	0.16	0.01
Gender										
Male	1,211 (66.10)	706 (64.70)	505 (68.1)	0.13	0.17	799 (63.80)	498 (63.60)	301 (64.20)	0.84	0.07
Female	622 (33.90)	385 (35.30)	237 (31.90)			453 (36.20)	285 (36.40)	168 (35.8)		
Occupation										
Retirees	858 (46.80)	462 (42.30)	396 (53.40)	<0.001	0.14	616 (49.20)	347 (44.30)	269 (57.40)	<0.001	0.04
Housework and unemployment	324 (17.70)	191 (17.50)	133 (17.90)			220 (17.60)	124 (15.80)	96 (20.50)		

Public servant	25 (1.40)	21 (1.90)	4 (0.50)			12 (1.00)	10 (1.30)	2 (0.40)		
Laborers	23 (1.30)	13(1.20)	10 (1.30)			13 (1.00)	7 (0.90)	6 (1.30)		
Cadres	35 (1.90)	24(2.20)	11 (1.50)			22 (1.80)	13 (1.70)	9 (1.90)		
Farmers and migrant workers	66 (3.60)	52 (4.80)	14 (1.90)			46 (3.70)	34 (4.30)	12 (2.60)		
Medical workers	21 (1.10)	14 (1.30)	7 (0.90)			14 (1.10)	8 (1.00)	6 (1.30)		
Other occupations	481 (26.20)	314 (28.80)	167 (22.50)			309 (24.70)	240 (30.70)	69 (14.70)		
Location										
Central area in Wuhan	1,385 (75.60)	818 (75.00)	567 (76.40)	0.08	0.16	922 (73.60)	569 (72.70)	353 (75.30)	0.23	0.10
Sub urban area in Wuhan	288 (15.70)	186 (17.00)	102 (13.70)			220 (17.60)	137 (17.50)	83 (17.70)		
Out of city/other	160 (8.70)	87 (8.00)	73 (9.80)			110 (8.80)	77 (9.80)	33 (7.00)		
Clinical characteristics										
Diabetes	357 (19.50)	117 (10.70)	240 (32.30)	<0.001	0.13	254 (20.30)	107 (13.70)	147 (31.30)	<0.001	0.04
Cardiovascular diseases	330 (18.00)	114 (10.40)	216 (29.10)	<0.001	0.15	251 (20.00)	107 (13.70)	144 (30.70)	<0.001	0.03
Cerebrovascular diseases	174 (9.50)	51 (4.70)	123 (16.60)	<0.001	0.04	140 (11.20)	57 (7.30)	83 (17.70)	<0.001	0.08
Respiratory	338 (18.40)	142 (13.00)	196 (26.40)	<0.001	0.26	268 (21.40)	148 (18.90)	120 (25.60)	0.005	0.04
Cancer	343 (18.70)	150 (13.70)	193 (26.00)	<0.001	0.11	266 (21.20)	153 (19.50)	113 (24.10)	0.041	0.07
Other diseases**	590 (32.20)	288 (26.40)	302 (40.70)	<0.001	0.10	447 (35.70)	262 (33.50)	185 (39.40)	0.032	0.05
# of underlying disease										
None	629 (34.30)	629 (57.70)	0 (0.00)	<0.001	0.04	373 (29.80)	373 (47.60)	0 (0.00)	<0.001	0.09
≤ 2	889 (48.50)	403 (36.90)	486 (65.50)			639 (51.00)	334 (42.70)	305 (65.00)		
>2	315 (17.20)	59 (5.40)	256 (34.50)			240 (19.20)	76 (9.70)	164 (35.00)		
COVID-19 severity* **										
Mild	597 (32.60)	418 (38.30)	179 (24.10)	<0.001	0.06	387 (30.90)	267 (34.10)	120 (25.60)	<0.001	0.04
Moderate	351 (19.10)	21 6(19.80)	135 (18.20)			247 (19.70)	232 (29.60)	15 (3.20)		

Severe	551 (30.10)	320 (29.30)	231 (31.10)			390 (31.20)	211 (26.90)	179 (38.20)		
Critical	334 (18.20)	137 (12.60)	197 (26.50)			228 (18.20)	73 (9.30)	155 (33.00)		
Date of onset										
Dec-1 to Dec-31-2019	33 (1.80)	20 (1.80)	13 (1.80)	0.08	0.19	21 (1.70)	15 (1.90)	6 (1.30)	<0.001	0.06
Jan-1 to Jan-15-2020	272 (14.80)	161 (14.80)	111 (150)			184 (14.70)	131 (16.70)	53 (11.30)		
Jan-16 to Jan-31-2020	1,159 (63.20)	672 (61.60)	487 (65.60)			787 (62.90)	450 (57.50)	337 (71.90)		
Feb-1 to Feb-15-2020	346 (18.90)	219 (20.10)	127 (17.10)			244 (19.50)	173 (22.10)	71 (15.10)		
Feb-16 to Feb-24-2020	23 (1.30)	19 (1.70)	4 (0.50)			16 (1.30)	14 (1.80)	2 (0.40)		
Date of diagnosis										
Dec-1 to Dec-31-2019	3 (0.20)	3 (0.30)	0 (0.00)	0.001	0.08	3 (0.20)	3 (0.40)	0 (0.00)	0.001	0.05
Jan-1 to Jan-15-2020	32 (1.70)	24 (2.20)	8 (1.10)			22 (1.8 0)	18 (2.30)	4 (0.90)		
Jan-16 to Jan-31-2020	509 (27.80)	288 (26.40)	221 (29.80)			254 (28.30)	229 (29.20)	125 (26.70)		
Feb-1 to Feb-15-2020	1157 (63.10)	679 (62.20)	478 (64.40)			788 (62.90)	466 (59.50)	322 (68.70)		
Feb-16 to Feb-24-2020	132 (7.20)	97 (8.90)	35 (4.70)			85 (6.80)	67 (8.60)	18 (3.80)		
Date of death										
Jan-1 to Jan-15-2020	1 (0.10)	1 (0.10)	0 (0.00)	<0.001	0.05	0 (0.00)	0 (0.00)	0 (0.00)	<0.001	0.03
Jan-16 to Jan-31-2020	223 (12.20)	144 (13.20)	79 (10.60)			158 (12.60)	119 (15.2)	39 (8.30)		
Feb-1 to Feb-15-2020	1,229 (67.00)	662 (60.70)	567 (76.40)			844 (67.40)	470 (60.00)	374 (79.70)		
Feb-16 to Feb-24-2020	380 (20.70)	284 (26)	96 (12.90)			250 (20.00)	194 (24.80)	56 (11.90)		

Duration from onset to diagnosis, Median (IQR)	10 (6, 14)	10 (6, 15)	10 (7, 14)	0.47	0.22	10 (6, 14)	10 (6, 14)	10 (7, 14)	0.85	0.10
Duration from onset to endpoint (death), Median (IQR)	17 (12, 22)	17 (12, 23)	15 (10, 21)	0.031	0.02	17 (11, 22)	17 (11, 23)	15 (10, 21)	0.047	0.03

**The Mann Whitney U-test was used to test the difference in the continuous variables and the Chi square or Fisher exact test was used to test the difference in the categorical variables between the two groups (with and without hypertension).*

***Other diseases included: anemia, hypothyroidism, Parkinson's disease, prostatic hyperplasia, fractures, etc.*

****The severity categories were according to the diagnostic criteria of the new coronavirus infection pneumonia diagnosis and treatment plan (trial fifth version).*

*****The standardized mean difference, defined as the mean difference between the groups divided by the standard deviation of the control (without hypertension) group.*

BOTTOM: association of hypertension and COVID-19 severity in the deceased COVID patients analyzed with an overall odds ratio (OR) higher in hypertensive patients by 2.4- to 2.9 ():

	Unmatched				**Matched**			
	Cases with hyper-tension/ all cases	**Uni-variate OR (95% CI)**	**Multi-variate OR (95% CI)***	**Multi-variate OR (95% CI)****	**Cases with hyper-tension/ all cases**	**Uni-vari-ate OR (95% CI)**	**Multi-variate OR (95% CI)***	**Multi-variate OR (95% CI)****
Mild CO-VID-19 illness	179/597	1.00 (refer-ence)	1.00 (refer-ence)	1.00 (refer-ence)	120/387	1.00 (refer-ence)	1.00 (refer-ence)	1.00 (refer-ence)
Moder-ate CO-VID-19 illness	135/351	2.62 (1.50–2.77)	3.40 (2.23–5.14)	2.80 (2.10–4.04)	15/247	1.02 (0.51–2.87)	1.14 (1.12–5.14)	2.60 (2.32–3.40)
Severe CO-VID-19 illness	231/551	4.72 (3.60–10.85)	9.19 (5.29–15.96)	11.66 (6.19–16.31)	179/390	2.72 (2.10–2.85)	9.13 (5.21–13.26)	10.60 (6.10–17.31)
Critical CO-VID-19 illness	197/334	9.44 (5.16–23.79)	42.76 (20.53–89.03)	36.16 (23.13–67.00)	155/228	2.64 (2.26–3.09)	32.16 (23.43–79.04)	35.02 (20.11–81.00)
P-trend		<0.001	<0.001	<0.001		<0.001	<0.001	<0.001
Non-severe CO-VID-19 illness***	314/948	1.00 (refer-ence)	1.00 (refer-ence)	1.00 (refer-ence)	135/634	1.00 (refer-ence)	1.00 (refer-ence)	1.00 (refer-ence)
Severe CO-VID-19 illness***	428/885	2.93 (2.62–3.04)	3.02 (2.56–3.10)	2.90 (2.00–3.10)	334/618	2.51 (2.17–2.91)	2.19 (2.09–2.76)	2.44 (1.77–3.08)

[a]*The adjustment for age and sex was conducted in all the multivariate models for the unmatched data, and the matching for age and sex was conducted in all the models for the matched analyses.*

**The estimated ORs (95% CIs) of hypertension after further adjustment for occupation and location.*

***The estimated ORs (95% CI) of hypertension after further adjustment for occupation, location and the other underlying diseases.*

****The non-severe COVID-19 illness included mild and moderate illnesses, while the severe COVOID-19 illness included severe and critical illnesses.*

A list of studies exploring the impact of hypertension on COVID-19 severity was compiled by Tadic, Saeed, Grassi, Taddei, Mancia and Cuspidi (2021) in *Frontiers in Cardiovascular Medicine* as presented below in table 9 ():

Reference	**Sample size**	**Age**	**Women (%)**	**Hypertension (%)**	**Other important findings**
Sun et al. (6)	3,400	61 (50–68)	1,751 (51)	1,782 (52)	Hypertension only together with diabetes was independent predictor of mortality and ARDS.
Bauer et al. (17)	1,449	54.7 ± 22.5	920 (63)	525 (36)	Hypertension was independent predictor of severe form of COVID-19 only in patients younger than 65 years, but not in the whole study population.
Barrera et al. (18)	15,794	–	–	–	Hypertension and diabetes separately were significant predictors of ICU and mortality, but not with severe COVID-19.
Grasselli et al. (19)	3,988	63 (56–69)	800 (20)	1,643 (41)	Diabetes and hypercholesterolemia, but not hypertension, were independent predictors of mortality in COVID-19 patients hospitalized in ICU.
Dennis et al. (21)	19,256	66	7,683 (40)	5,657 (29)	Patients with type 2 diabetes were at increased risk of mortality independently of hypertension, chronic respiratory disease, chronic heart disease, chronic renal disease, chronic liver disease and other potential risk factors.
Gupta et al. (16)	2,215	60.5 ± 14.5	779 (35)	1,322 (60)	Body mass index ≥ 40 kg/m^2 and coronary artery disease, but not hypertension and diabetes, were independent predictors of 28-day mortality in COVID-19 patients.

ARDS, acute respiratory distress syndrome; ICU, intensive care unit.

5.3 PERSPECTIVES OF HYPERTENSIVE PATIENTS

Godinas et al (2021) in the *Orphanet Journal of Rare Diseases* discussed the PH-CARE-COVID survey aimed at gauging the perspectives of pulmonary hypertension (PH) patients during the COVID-19 pandemic. This was done keeping in mind the shift from regular care these patients needed to COVID-19 care (mainly as directed by authorities). This survey was spearheaded by the following: "*European Pulmonary Hypertension Association (PHA Europe), the European Reference Network for rare lung diseases (ERN-LUNG), the European Respiratory Society (ERS) Assembly 13 on Pulmonary Vascular Diseases, the ERS Clinical Research Collaboration PHAROS, the European Lung Foundation (ELF) and the European Society of Cardiology (ESC) Working Group on Pulmonary Circulation & Right Ventricular Function*". The survey (in 16 languages) was inclusive of looking at the perspectives of the patients.

73% of 1073 participants from 52 countries were hypertensive patients with the balance accounted for by relatives. 96% of the sample was adult. Table 10 below ***TOP***: depicts the country-wise list of participants (Godinas et al, 2021: *Open Access: This article is licensed under a Creative Commons Attribution 4.0 International License, which permits use, sharing, adaptation, distribution and reproduction in any medium or format, as long as you give appropriate credit to the original author(s) and the source, provide a link to the Creative Commons licence, and indicate if changes were made. The images or other third party material in this article are included in the article's Creative Commons licence, unless indicated otherwise in a credit line to the material. If material is not included in the article's Creative Commons licence and your intended use is not permitted by statutory regulation or exceeds the permitted use, you will need to obtain permission directly from the copyright holder. To view a copy of this licence, visit http://creativecommons.org/licenses/by/4.0/. The Creative Commons Public Domain Dedication waiver (http://creativecommons.org/publicdomain/zero/1.0/) applies to the data made available in this article, unless otherwise stated in a credit line to the data*):

Name of country	**Total answers per country**	**% of the total answers of the survey**
Belgium	144	13.4
France	125	11.6
The Netherlands	116	10.8
Chile	90	8.4
Spain	78	7.3

Ukraine	74	7.0
Portugal	66	6.2
Italy	47	4.4
Austria	37	3.4
Germany	29	2.7
Serbia	24	2.2
Latvia	23	2.1
Slovakia	21	2.0
Slovenia	18	1.7
Canada	17	1.6
India	17	1.6
United Kingdom	15	1.4
Brazil	13	1.2
Colombia	12	1.1
Czechia	12	1.1
USA	9	0.8
Bosnia	9	0.8
Croatia	9	0.8
Lithuania	8	0.7
Argentina	7	0.7
Peru	7	0.7
Uruguay	6	0.6
Australia	4	0.4
Bulgaria	4	0.4
Finland	3	0.3
Dominican Republic	3	0.3
Norway	2	0.2
Poland	2	0.2
Sweden	2	0.2
Honduras	2	0.2
Greece	2	0.2
New Zealand	1	<0.1
Mexico	1	<0.1
Montenegro	1	<0.1
El Salvador	1	<0.1
Costa Rica	1	<0.1
Ecuador	1	<0.1
Venezuela	1	<0.1

Switzerland	1	<0.1
Algeria	1	<0.1
Morocco	1	<0.1
Malaysia	1	<0.1
South Africa	1	<0.1
Russia	1	<0.1
Hungary	1	<0.1
Lebanon	1	<0.1
Tajikistan	1	<0.1

BOTTOM: details of hypertension and therapy (Godinas et al, 2021: *Open Access: This article is licensed under a Creative Commons Attribution 4.0 International License, which permits use, sharing, adaptation, distribution and reproduction in any medium or format, as long as you give appropriate credit to the original author(s) and the source, provide a link to the Creative Commons licence, and indicate if changes were made. The images or other third party material in this article are included in the article's Creative Commons licence, unless indicated otherwise in a credit line to the material. If material is not included in the article's Creative Commons licence and your intended use is not permitted by statutory regulation or exceeds the permitted use, you will need to obtain permission directly from the copyright holder. To view a copy of this licence, visit http://creativecommons.org/licenses/by/4.0/. The Creative Commons Public Domain Dedication waiver (http://creativecommons.org/publicdomain/zero/1.0/) applies to the data made available in this article, unless otherwise stated in a credit line to the data*):

N=1073	**N**
Female	839
Male	228
No answer	6
Idiopathic PAH	416
Heritable/genetic PAH	80
Drug related PAH	15
Liver disease related PAH	16
Connective tissue disease related PAH	93
Congenital heart malformation related PAH	208

Chronic thromboembolic PH	161
Other	13
I do not know/I am not sure/Do not want to answer	71
Oral	934
Intravenous	76
Subcutaneous	77
Inhaled	72
Study medication	18
No specific PH treatment reported	115

While 47% of the patients demonstrated no new symptoms, the predominant symptoms were on the lines of PH or viral diseases with 3% reporting a loss of smell or taste. 76% of the patients reported stability in the physical conditions with 14% reporting deterioration (more than 50% of them ascribed this to PH).

Figure 7 below is representative of ***TOP***: a slew of symptoms reported by the PH patients: ENT: ear, nose and throat symptoms (Godinas et al, 2021: *Open Access: This article is licensed under a Creative Commons Attribution 4.0 International License, which permits use, sharing, adaptation, distribution and reproduction in any medium or format, as long as you give appropriate credit to the original author(s) and the source, provide a link to the Creative Commons licence, and indicate if changes were made. The images or other third party material in this article are included in the article's Creative Commons licence, unless indicated otherwise in a credit line to the material. If material is not included in the article's Creative Commons licence and your intended use is not permitted by statutory regulation or exceeds the permitted use, you will need to obtain permission directly from the copyright holder. To view a copy of this licence, visit http://creativecommons.org/licenses/by/4.0/. The Creative Commons Public Domain Dedication waiver (http://creativecommons.org/publicdomain/zero/1.0/) applies to the data made available in this article, unless otherwise stated in a credit line to the data*):

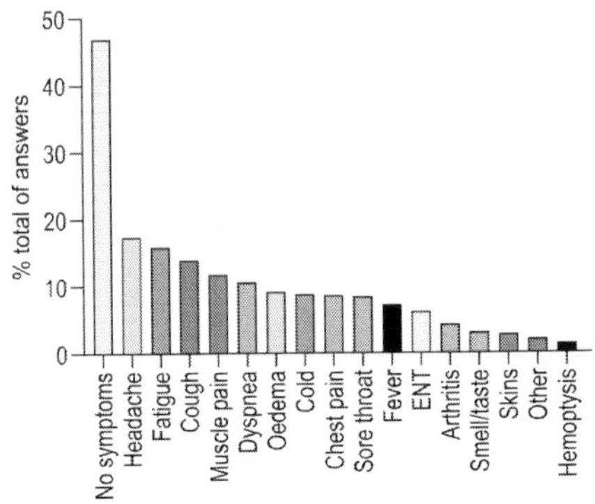

BOTTOM: a) self-reported health and b) reasons ascribed for deterioration of health (Godinas et al, 2021: *Open Access: This article is licensed under a Creative Commons Attribution 4.0 International License, which permits use, sharing, adaptation, distribution and reproduction in any medium or format, as long as you give appropriate credit to the original author(s) and the source, provide a link to the Creative Commons licence, and indicate if changes were made. The images or other third party material in this article are included in the article's Creative Commons licence, unless indicated otherwise in a credit line to the material. If material is not included in the article's Creative Commons licence and your intended use is not permitted by statutory regulation or exceeds the permitted use, you will need to obtain permission directly from the copyright holder. To view a copy of this licence, visit http://creativecommons.org/licenses/by/4.0/. The Creative Commons Public Domain Dedication waiver (http://creativecommons.org/publicdomain/zero/1.0/) applies to the data made available in this article, unless otherwise stated in a credit line to the data*):

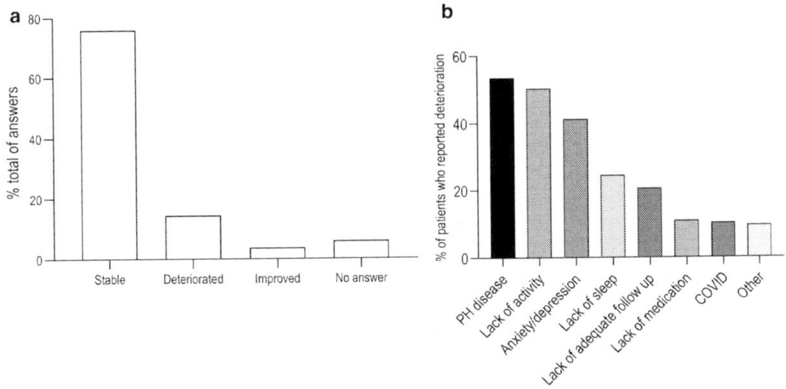

Interestingly, only 10% of these PH patients underwent testing for COVID-19 with 90% of these emerging negative. Only 1% was admitted to a hospital for COVID-19.

For issues/questions concerning PH, 44% of them were in touch with health caregivers (GP, PH physician, special nurse, cardiologist and emergency care). 59% of the participants reported no contact with PH specialists due to mainly fear of visiting a hospital followed by fear of disturbing. 64% reported cancellation of an appointment from their PH center while 16% of them made the cancellation themselves.

Figure 8 below is illustrative of how the patients felt during the lockdown (Godinas et al, 2021: *Open Access: This article is licensed under a Creative Commons Attribution 4.0 International License, which permits use, sharing, adaptation, distribution and reproduction in any medium or format, as long as you give appropriate credit to the original author(s) and the source, provide a link to the Creative Commons licence, and indicate if changes were made. The images or other third party material in this article are included in the article's Creative Commons licence, unless indicated otherwise in a credit line to the material. If material is not included in the article's Creative Commons licence and your intended use is not permitted by statutory regulation or exceeds the permitted use, you will need to obtain permission directly from the copyright holder. To view a copy of this licence, visit http://creativecommons.org/licenses/by/4.0/. The Creative Commons Public Domain Dedication waiver (http://creativecommons.org/publicdomain/zero/1.0/) applies to the data made available in this article, unless otherwise stated in a credit line to the data*):

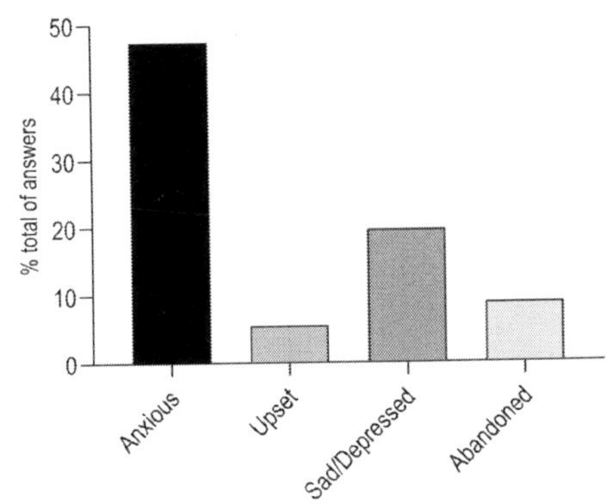

Given that 12% of PH patients established direct contact with their center, 42% contacted via phone followed by video call, email and post, the avenue of telemedicine looks promising that warrants appropriate planning for its incorporation during such exigencies.

Final Points

The presence of ***comorbidities*** is a challenge for COVID-19 patients. According to Flaherty and team (2020) in *Tropical Diseases, Travel Medicine and Vaccines*, the ***impact of COVID-19*** is overall summarized below in table 11 (Flaherty et al, 2020: *Open Access: This article is licensed under a Creative Commons Attribution 4.0 International License, which permits use, sharing, adaptation, distribution and reproduction in any medium or format, as long as you give appropriate credit to the original author(s) and the source, provide a link to the Creative Commons licence, and indicate if changes were made. The images or other third party material in this article are included in the article's Creative Commons licence, unless indicated otherwise in a credit line to the material. If material is not included in the article's Creative Commons licence and your intended use is not permitted by statutory regulation or exceeds the permitted use, you will need to obtain permission directly from the copyright holder. To view a copy of this licence, visit http://creativecommons.org/licenses/by/4.0/. The Creative Commons Public Domain Dedication waiver (http://creativecommons.org/publicdomain/zero/1.0/) applies to the data made available in this article, unless otherwise stated in a credit line to the data*):

Key: *ICU* Intensive care unit, *COPD* Chronic obstructive pulmonary disease, *RAAS* Renin-angiotensin-aldosterone system

Health issue	Comorbidities	Impact of COVID-19
Predisposition to severe disease	Obesity	More severe symptoms of COVID-19; requirement for complex ICU management; poorer clinical outcomes
	Diabetes	Increased clinical severity and mortality
Access to routine medical care	COPD	Cancellation of smoking cessation clinics
	Cystic fibrosis	Decreased access to multidisciplinary care and transplantation
	Cancer	Necessity to defer chemotherapy in some cases
	Chronic kidney disease	Infection risk in dialysis units; decreased access to renal transplantation
	Mental health disorders	Barriers to accessing outpatient care

Influence of chronic medications	Hypertension	Controversy around use of RAAS blockers
	Autoimmune disease	Immunosuppressant drugs predispose to infection
	Chronic kidney disease	Immunosuppressant drugs predispose to infection
Negative effects of pandemic lockdown	Obesity	Restrictions on physical activity; excessive caloric intake
	Mental health disorders	Social isolation; heightened anxiety; sleep disturbance; reduced capacity of psychotic patients to adhere to public health advice

Hosseinzadeh and team (2021) explored looking at ***prognostic factors*** in COVID-19 patients with hypertension by scrutinizing at clinical, lab and epidemiological levels. This retrospective study published in *Clinical Hypertension* looked at the medical profiles of 598 COVID-19 patients in Baqiyatallah Hospital in Tehran, Iran. Hypertension was seen in 29.4% of the patients while 40.9% of these patients were also diabetic. The hypertensive group demonstrated an elevated profile of renal disorders and cardiovascular disease. Further, the group with hypertension displayed more critical and severe COVID-19 vs. normotensive patients.

Table 12 below documents these features in hypertensive patients with COVID-19 vs. normotensive ones : *MI: myocardial infarction, CVD: cardiovascular disease* (Hosseinzadeh et al, 2021: *Open Access: This article is licensed under a Creative Commons Attribution 4.0 International License, which permits use, sharing, adaptation, distribution and reproduction in any medium or format, as long as you give appropriate credit to the original author(s) and the source, provide a link to the Creative Commons licence, and indicate if changes were made. The images or other third party material in this article are included in the article's Creative Commons licence, unless indicated otherwise in a credit line to the material. If material is not included in the article's Creative Commons licence and your intended use is not permitted by statutory regulation or exceeds the permitted use, you will need to obtain permission directly from the copyright holder. To view a copy of this licence, visit http://creativecommons.org/licenses/by/4.0/. The Creative Commons Public Domain Dedication waiver (http://creativecommons.org/*

publicdomain/zero/1.0/) applies to the data made available in this article, unless otherwise stated in a credit line to the data):

Variable	**Hypertensive (N: 176)**	**Non-hypertensive (N:422)**	***P*-value**
Age (Year)	58.21±22.5	56.15±20.2	0.18
BMI (kg/m^2)	28.62±4.6	28.01±4.3	0.13
Sex (male)	115(65.3%)	289(68.5%)	0.78
Education			0.65
• < diploma	57 (32.4%)	111(26.3%)	
• >Diploma	119(67.6%)	311 (73.7%)	
Exposure history	30(17%)	191 (45.2%)	0.45
• Contact with patients	118(67%)	250 (59.2%)	0.061
• Go hospital	30(17%)	87(20.6%)	0.23
• Public transportation	24(22.5%)	118(24.1%)	0.64
Comorbidities (positive)	108(61.3%)	249(59%)	0.75
Current smoker	4(2.2%)	12(2.8%)	0.92
Diabetes	72(40.9%)	76(18%)	0.001
CVD	63(35.7%)	44(10.3%)	0.001
MI	4(2.3%)	2(0. 5%)	0.061
Lung diseases	36(20.4%)	52(12.3%)	0.07
Asthma and Allergy	31(17.6%)	67 (15.9%)	0.45
Kidney diseases	25(14.3%)	33(7.8%)	0.013
Liver diseases	10(5.7%)	22(5.2%)	0.47
Cancer	4(2.3%)	8(1.9%)	0.68
Blood Group			
• A	55(31.3%)	139(32.8%)	0.21
• B	45(25.5%)	110(26.3%)	0.71
• AB	21(11.9%)	60(14.2%)	0.52
• O	55(31.3%)	113(26.7%)	0.31
Clinical type of COVID- 19 infection on admission			
Severe/Critical type (%)	42(23.8%)	41(9.7%)	0.012

Further, 44.3% of the hypertensive patients showed high fever (> 38.5) vs. 9.6% in the normotensive group. A similar trend was also observed for dyspnea (shortness of breath). A statistical significance was also observed for the longer duration of hospital stay in 22.9% of hypertensive patients as opposed to the normotensive ones. Table 13 below is demonstrative of these observations among others : *PCR: Polymerase chain reaction, ALT: Alanine aminotransferase AST: Aspartate transaminase, BS: blood sugar, INR: International normalized ratio, CRP: c-reactive protein* (Hosseinzadeh et al, 2021: *Open Access: This article is licensed under a Creative Commons Attribution 4.0 International License, which permits use, sharing, adaptation, distribution and reproduction in any medium or format, as long as you give appropriate credit to the original author(s) and the source, provide a link to the Creative Commons licence, and indicate if changes were made. The images or other third party material in this article are included in the article's Creative Commons licence, unless indicated otherwise in a credit line to the material. If material is not included in the article's Creative Commons licence and your intended use is not permitted by statutory regulation or exceeds the permitted use, you will need to obtain permission directly from the copyright holder. To view a copy of this licence, visit http://creativecommons.org/licenses/by/4.0/. The Creative Commons Public Domain Dedication waiver (http://creativecommons.org/publicdomain/zero/1.0/) applies to the data made available in this article, unless otherwise stated in a credit line to the data*):

Symptoms	**Normal range**	**Hyperten-sive (N: 176)**	**Non-hy-pertensive (N:422)**	**P value**
Fever	N (%)			0.042
• ≥38.5 c		78(44.3%)	125(29.6%)	
• <38.5		98(45.7%)	300(70.4%)	
Cough	N (%)	103 (58.5%)	257(60.9%)	0.38
Dysphonia	N (%)	105(59.6%)	261(61.8%)	0.37
Chest pain	N (%)	38(21.6%)	107(25.3%)	0.19
Dizziness	N (%)	19(10.7%)	56(13.2%)	0.18
Headache	N (%)	56(31.8%)	153(36.2%)	0.18
Weakness	N (%)	90(51.4%)	232(55.1%)	0.24
Rhinorrhea	N (%)	20(11.3%)	41(9.7%)	0.3
Myalgia	N (%)	62(35.2%)	154(36.5%)	0.61

Test Sense	N (%)	21(11.9%)	51(12%)	0.91
Shortness of breath	N (%)	22(12.5%)	20(4.7%)	0.023
Diarrhea	N (%)	34(19.3%)	84(19.9%)	0.76
Vomiting	N (%)	53(30.2%)	136(32.3%)	0.23
Laboratory tests				
Leucocytes (× 109/L)	4–10	7.61±7.45	6.26±3.08	0.14
Neutrophils (×109/L)	2–7	6.28±0.65	5.23±.065	0.12
Lymphocytes (×109/L)	0.8–4	2.09±1.68	2.06±1.68	0.31
Platelets (×109/L)	125–450	164.2±63.1	159.6±75.2	0.12
Hemoglobin (g/L)	115–150	15.67±2.7	14.8±2.8	0.068
Hematocrit (%;)	33.35–50.8	44.15±3.7	42.1±4.3	0.23
Albumin (g/L)	40–55	44.15±7.8	41.5±6.5	0.063
Alt(U/L)	9–50	40.5±24.6	39.6±29.2	0.081
Ast(U/L)	15–40	36.1±26.2	32.2±25.1	0.038
Total bilirubin)	0–26	9.1±3.1	8.6±4.2	0.25
Serum sodium (mmol/L)	137–147	135.6±15.5	134.1±18.1	0.71
Serum potassium (mmol/L)	3.5–5.3	3.41±2.4	3.31±2.9	0.77
Glucose (mmol/L)	70–140	5.99±2.15	5.01±1.96	0.62
INR	0.85–1.15	1.41±0.17	1.15±0.13	0.28
CRP (mg/L) d	0–10	41.2±32.1	39.8±31.2	0.11
Chest X-ray/CT findings	N (%)			
• Normal		19(10.7%)	50(11.8%)	0.24
• Unilateral pneumonia		25(14.2%)	86(20.3%)	0.046
• Bilateral pneumonia		68 (38.6%)	156(36.9%)	0.12
• Multiple mottling and ground-glass opacity		64(36.5%)	159(31%)	0.071
Clinical outcome				

Increased hospital stays	N (%)	41(22.9%)	46(10.9%)	0.015
Death	N (%)	16(9.1%)	25(5.9%)	0.11
Readmission	N (%)	20(11.3%)	42(9.9%)	0.28

While hypertension is indeed a risk factor for severe disease, as documented by research, other factors were also unearthed inclusive of BMI, increased age, diabetes, CVD and chronic kidney disease as shown below in table 14 (Hosseinzadeh et al, 2021: *Open Access: This article is licensed under a Creative Commons Attribution 4.0 International License, which permits use, sharing, adaptation, distribution and reproduction in any medium or format, as long as you give appropriate credit to the original author(s) and the source, provide a link to the Creative Commons licence, and indicate if changes were made. The images or other third party material in this article are included in the article's Creative Commons licence, unless indicated otherwise in a credit line to the material. If material is not included in the article's Creative Commons licence and your intended use is not permitted by statutory regulation or exceeds the permitted use, you will need to obtain permission directly from the copyright holder. To view a copy of this licence, visit http://creativecommons.org/licenses/by/4.0/. The Creative Commons Public Domain Dedication waiver (http://creativecommons.org/publicdomain/zero/1.0/) applies to the data made available in this article, unless otherwise stated in a credit line to the data*):

Variable	**Group**					
	Non-hypertensive			**Non-hypertensive**		
	O R Adj	**95% CI**	**P value**	**O R Adj**	**95% CI**	**P value**
Age>60year-old	1.26	1.08-1.46	0.021	1.06	0 1 1 . -2.08	0.044
BMI>25(kg/m2)	1.8	1.02–2.42	0.027	1.34	0.88–1.89	0.21
Increased hospital stays (positive)	2.1	1.24–2.97	0.013	1.48	0.36–2.61	0.28
CVD (positive)	3.3	1 . 3 – 5.4	0.001	2.8	0.84–4.78	0.058
Diabetes (positive)	2.22	1.15–3.31	0.001	1.24	0.59–1.92	0.28
Kidney daisies (positive)	1.83	1.19–2.21	0.013	1.22	0.66–1.81	

Christopher E. Clark and team (2021) in the *Journal of Human Hypertension* in "*A scientific statement on behalf of the British and Irish Hypertension Society*" reported that strong recommendations become challenging given the gaps in data. They stated that prolonged and poorly controlled hypertension would be a higher risk that warrants the practicing of social distancing recommendations. Further, the safety and importance of current medication are essential to be conveyed to hypertensive patients. The final recommendations are inclusive of correct medication adherence and relevant lifestyle practices.

5.4 REFERENCES

1. De Lorenzo R, Conte C, Lanzani C, Benedetti F, Roveri L, Mazza MG, et al. (2020) Residual clinical damage after COVID-19: A retrospective and prospective observational cohort study. PLoS ONE 15(10): e0239570. https://doi.org/10.1371/journal.pone.0239570
2. Moazzami B, Chaichian S, Kasaeian A, Djalalinia S, Akhlaghdoust M, Eslami M, et al. (2020) Metabolic risk factors and risk of Covid-19: A systematic review and meta-analysis. PLoS ONE 15(12): e0243600. https://doi.org/10.1371/journal.pone.0243600
3. de Almeida-Pititto, B., Dualib, P.M., Zajdenverg, L. *et al.* Severity and mortality of COVID 19 in patients with diabetes, hypertension and cardiovascular disease: a meta-analysis. *Diabetol Metab Syndr* **12,** 75 (2020). https://doi.org/10.1186/s13098-020-00586-4
4. Bajgain, K. T., Badal, S., Bajgain, B. B., & Santana, M. J. (2021). Prevalence of comorbidities among individuals with COVID-19: A rapid review of current literature. *American journal of infection control*, *49*(2), 238–246. https://doi.org/10.1016/j.ajic.2020.06.213
5. Emami, A., Javanmardi, F., Pirbonyeh, N., & Akbari, A. (2020). Prevalence of Underlying Diseases in Hospitalized Patients with COVID-19: a Systematic Review and Meta-Analysis. *Archives of academic emergency medicine*, *8*(1), e35.
6. Chen G, Li X, Gong Z, Xia H, Wang Y, Wang X, et al. (2021) Hypertension as a sequela in patients of SARS-CoV-2 infection. PLoS ONE 16(4): e0250815. https://doi.org/10.1371/journal.pone.0250815
7. Osikomaiya, B., Erinoso, O., Wright, K.O. *et al.* 'Long COVID': persistent COVID-19 symptoms in survivors managed in Lagos State, Nigeria. *BMC Infect Dis* **21,** 304 (2021). https://doi.org/10.1186/s12879-020-05716-x
8. Mubarik S, Liu X, Eshak ES, Liu K, Liu Q, Wang F, Shi F, Wen H, Bai J, Yu C and Cao J (2021) The Association of Hypertension With the Severity of and Mortality From the COVID-19 in the Early Stage of the Epidemic in Wuhan, China: A Multicenter Retrospective Cohort Study. *Front. Med.* 8:623608. doi: 10.3389/fmed.2021.623608
9. Tadic M, Saeed S, Grassi G, Taddei S, Mancia G and Cuspidi C (2021) Hypertension and COVID-19: Ongoing Controversies. *Front. Cardiovasc. Med.* 8:639222. doi: 10.3389/fcvm.2021.639222

10. Godinas, L., Iyer, K., Meszaros, G. *et al.* PH CARE COVID survey: an international patient survey on the care for pulmonary hypertension patients during the early phase of the COVID-19 pandemic. *Orphanet J Rare Dis* **16,** 196 (2021). https://doi.org/10.1186/s13023-021-01752-1
11. Flaherty, G.T., Hession, P., Liew, C.H. *et al.* COVID-19 in adult patients with pre-existing chronic cardiac, respiratory and metabolic disease: a critical literature review with clinical recommendations. *Trop Dis Travel Med Vaccines* **6,** 16 (2020). https://doi.org/10.1186/s40794-020-00118-y
12. Hosseinzadeh, R., Goharrizi, M.A.S.B., Bahardoust, M. *et al.* Should all patients with hypertension be worried about developing severe coronavirus disease 2019 (COVID-19)?. *Clin Hypertens* **27,** 3 (2021). https://doi.org/10.1186/s40885-021-00161-7
13. Clark, C.E., McDonagh, S.T.J., McManus, R.J. *et al.* COVID-19 and hypertension: risks and management. A scientific statement on behalf of the British and Irish Hypertension Society. *J Hum Hypertens* **35,** 304–307 (2021). https://doi.org/10.1038/s41371-020-00451-x

INDEX

A

acoustic radiation 151
adolescents 47, 48, 53, 54, 58, 61, 64, 65, 69, 80, 81, 89, 92, 95, 97, 104, 105, 107, 108, 109
Ambulatory Blood Pressure Monitoring 51
ambulatory BP measurements (ABPM) 5
animals 134
anterior thalamic radiation 151
anxiety 166, 170, 172, 174, 176, 178, 186, 187, 191, 201, 204

B

Beck Anxiety Inventory (BAI) 166
Beck Depression Inventory (BDI) 166
bilateral anterior 150
blood pressure 6, 7, 8, 9, 10, 21, 33, 34, 37, 38, 39, 40, 44
Brazilian Cardiovascular Adolescent Health (BRACAH) 97

C

calcium 112, 119, 125, 126, 127, 128
Carbohydrates 112
cardiovascular disease 224, 227, 230, 231, 255, 261
causative agent 220
children 47, 48, 51, 53, 54, 58, 61, 62, 64, 65, 68, 69, 72, 75, 81, 85, 92, 95, 96, 97, 103, 104, 105, 107, 108, 109
China Health and Retirement Longitudinal Study (CHARLS) 14
cingulum 149
cingulum bundle 150
Clinical Practice Guidelines (CPG) 60
Collaborative Perinatal Project (CPP) 86
conjugated bilirubin (CB) 86
COPDs ((chronic obstructive pulmonary diseases) 230

Coronary Artery Risk Development in Young Adults (CARDIA) 164
Coronavirus disease-2019 (COVID-19) 220
coupling strength (CS) 155

D

dairy products 112, 114, 115, 116, 117
depression 145, 166, 167, 170, 172, 174, 175, 176, 178, 179, 180, 186, 187, 188, 189, 191, 193, 194, 198, 199, 200, 201, 203, 204, 205, 217, 218
diabetes 222, 223, 224, 227, 230, 231, 236, 238, 259, 261
Diabetology & Metabolic Syndrome 224
diastolic blood pressure (DBP) 6, 65
diastolic BP (DBP) 53, 95
Dietary Approaches to Stop Hypertension (DASH) 112
Dietary Approaches to Stop Hypertension (DASH) diet 112
diffusion tensor imaging (DTI) 149
Digit Symbol Substitution Test (DSST) 164

E

education 2, 13, 17, 24
effective connectivity (EC) 155
Emergency Department (ED) 221
endothelial cells 80
essential hypertension (EH) 134
European Lung Foundation (ELF) 248
European Respiratory Society (ERS) 247
European society of cardiology (ESC) 7
European society of hypertension (ESH) 7

F

food 112, 133
fructose consumption 48, 85
functional connectivity (FC) 155, 156
functional magnetic resonance imaging (fMRI) 154

G

Generalized Health Anxiety Disorder 174
gestational hypertension (GH) 154
global health 3
global prevalence 48
grey matter volume (GMV) 149

H

healthy control (HC) 154
Healthy Lifestyle in Europe by Nutrition in Adolescence (HELENA) 97
High Blood Pressure 48, 60, 61
high-income countries (HICs) 18
home BP measurements (HBPM) 5
hormonal changes 65
Human development index (HDI) 2
hypertension 2, 3, 4, 5, 6, 7, 8, 10, 11, 12, 13, 14, 16, 17, 18, 20, 21, 24, 26, 27, 28, 29, 32, 33, 34, 36, 37, 40, 41, 42, 43, 44
hypertensive patients 151, 155, 175, 193, 194, 210, 211

I

inferior longitudinal fasciculus 149, 151
in sustained hypertension (SH) 11

K

kidney disease 230, 254, 259

L

lower-middle-income countries (lower-MICs) 18
low-income countries (LICs) 18

M

magnesium 112, 119, 130, 131, 132
malnourishment 221
masked hypertension (MH) 11
meat 112, 114, 115, 133
medication 151, 164, 173, 179, 201, 203, 204, 210, 214
Mendelian randomization (MR) 163
mental health 221
meta-analyses 119
metabolic risk factors 222
micronutrients 112, 119, 133
MicroRNAs (miRNAs) 134
Mini-Mental State Examination (MMSE) 157
Mini-Mental State Exam (MMSE) 146
Mini Nutritional Assessment (MNA) 220
modified Medical Research Council (mMRC) 220
Montreal Cognitive Assessment (MoCA) 157

N

National Health and Nutrition Examination Survey (NHANES) 53
National Institute of Health (NIH) 112
near-infrared spectroscopy (NIRS) 155
neurite orientation dispersion and density imaging (NODDI) 149
nitric oxide 80
Nuts 112, 133

O

obese (OB) 70
obesity 48, 74, 75, 89, 104
overweight (OW) 70

P

Patient Health Questionnaire 174, 179, 188, 193
patients 220, 221, 223, 226, 229, 230, 231, 232, 233, 234, 235, 238, 241, 242, 245, 247, 248, 251, 252, 253, 255, 256, 260, 261, 262
physical activity (PA) 97
plants 134
posterior thalamic radiation 150, 151
potassium 112, 119, 128
Preferred Reporting Items for Systematic Reviews and Meta-Analyses (PRISMA) 222
processed foods 112
pulmonary arterial hypertension (PAH) 166, 186

pulmonary Hypertension (PH) 174
pulse pressures 146

R

regional global connectivity (rGC) 154
renal vasoconstriction 80
renin–angiotensin system 80
Rey Auditory Verbal Learning Test (RAVLT) 164

S

Scatter plots 4
sedentary behavior (SB) 97
seeds 112, 113, 114, 115, 133
serum uric acid 80
Severe Acute Respiratory Syndrome-Coronavirus-2 (SARS-CoV-2) 220
Shanghai children and adolescents disease risk factor study\" (SCADRFS) 69
short-chain fatty acids (SCFAs) 135
smoking 230, 254
snapshot 48, 104
socioeconomic status (SES) 13
South African Stress and Health (SASH) 170
sugar-sweetened beverages (SSBs) 81
superior longitudinal fasciculus 150, 151, 154
systolic blood pressure (SBP) 6, 65
systolic BP (SBP) 53, 95

T

the gut-liver axis (GLA) 133
total brain volume (TBV) 149
total serum bilirubin (TSB) 86
Trail Making Part A (TMT-A) 146
Trail Making Part B (TMT-B) 146

U

unconjugated bilirubin (UCB) 86
United Nations Development Programme (UNDP) 2
upper-middle-income countries (upper-MICs) 18
uric acid 80, 85, 109

V

vegetables 112, 113, 114, 115, 116, 117
visuo-analog scale (VAS) 220

W

waist circumference (WC) 70
wavelet phase coherence (WPCO) 155
white-coat hypertension (WCH) 11
white matter volume (WMV) 149
Women's Health and Aging Study (WHAS) 146